Natural Resources, S

Angela Mendonca • Ana Cunha
Ranjan Chakrabarti
Editors

Natural Resources, Sustainability and Humanity

A Comprehensive View

 Springer

Editors
Angela Mendonca
Lamacaes School Cluster
Braga
Portugal

Ana Cunha
Department of Biology
School of Sciences
Minho University
Braga
Portugal

Ranjan Chakrabarti
Department of History
Jadavpur University
Kolkata
India

Disclaimer: The facts and opinions expressed in this work are those of the editors and authors and not necessarily those of the publisher.

ISBN 978-94-007-9602-7   ISBN 978-94-007-1321-5 (eBook)
DOI 10.1007/978-94-007-1321-5
Springer Dordrecht Heidelberg London New York

Springer is part of Springer Science+Business Media (www.springer.com)

Dedication

This book is dedicated to all anthropogenic climate changes casualties.

And to Prof Dr Hugh Freeman who died in May 2011. He was an internationally renowned psychiatrist whose major interest was unique and humane in concern with the effect of the environment on mental health. He presented a paper on this subject in Braga, Portugal, at the II International School Congress in May 2010. Prof Hugh Freeman was Editor of the British Journal of Psychiatry and published very many books including Psychiatric Cultures Compared, A Century of Psychiatry and The Impact of the Environment on Mental Disorder.

This is the abstract of the very last conference given by our dearest friend Hugh Freeman.

Mental Health Effects of the Environment
Professor Hugh Freeman
Green-Templeton College, Oxford 0X2 6HG, UK

At the same time as the recognised major changes in climate and the chemical composition of the atmosphere, there have also been upheavals in people's personal worlds. There have been large-scale movements of populations around the world. It is well known that changes in physical environments affect people's mental health and their behaviour. These might be factors such as overcrowding and facing novel interactions of different social and racial groups.

The biggest of these social changes is the drift to the cities, on a scale never known before in human history. For the first time, more people in the world now live in urban rather than rural settlements. Many cities, particularly in the Developing World, are of a far greater size than at any other time. This urbanisation has been completely unplanned and there is still not enough reliable understanding about its effects on people, notably any psychiatric changes caused by these movements. But there is some indication, for example, that the pressure of overcrowding could be a risk factor for the development of schizophrenia. It is time for concern about the effects of environmental changes on human health to include the psychiatric and wider emotional consequences.

This tree was planted in honour of Hugh Freeman at Peneda-Gêres Nacional Park, Portugal—In nature we trust.

Preface

Some books are meant to be read from the end to the beginning.

This is one of those "out of the box", you may choose wherever to start.

There will be no "how to read or how to go through". The chapters have been aligned as if visiting places and contemplating landscapes at ones pleasure.

That is one of the reasons this introduction is an attempt to answer Zenita Guenther's last chapter capital question: "Who ties the knot?"

This preface is no more than the way the book ends too. It raises questions and is supposed to let you unsettled, you have the best brain in the planet—use it, please.

The 2nd International School Congress: Natural Resources, Sustainability and Humanity took place in Braga, Portugal from the 5th to the 9th of May, 2010.

We have started working three years in advance, focusing in the articulation of the axis Environment, Humanity and School needs, and adaptation to a changing world.

The first two because from them emanate the largest human concerns, the other because it is a privileged locus of generational transformation and a tool to hinder, stop or find new solutions.

We have aimed to unite the educational communities of Portugal and worldwide education communities that wanted to join us, in an environment of formation and information, driven and challenged by the words of scientists, researchers, educators, politicians, entrepreneurs, artists and writers which have demonstrated more availability, enthusiasm, hence the most motivated and qualified ones to work at our assembly of speakers.

The event had three major meetings aiming different target audiences and outcomes:

From the 5th to the 6th, we had gathered in a Small Meeting, where solutions and mitigations proposals were discussed for social, cultural, environmental and educational issues.

Then, from the 6th to the 9th, we had gathered at a Large Meeting where we worked for the (in)formation and awareness of educators and the general public. We shall be facing different and diverse points of view, experiences and backgrounds, but we shall be focused in the holistic construction of human beings, aware and capable of making the future a better place.

Amongst this dates we launched for the first time in Portugal the 1st International Workshop on the History of the Environment: *On the History of Environment and Global Climate Change—Water, Ecology, De-forestation, Agriculture, Politics, and the Management of Nature.*

This event's philosophy had as a starting point the I International School Congress: Environment, Health and Education (I ISC) that took place from the 8th to the 10th of May, 2008 in Braga. Although we may have considered the previous event a success in several aspects, we dared to expand the achieved objectives bringing more specific themes to the public discussion. We continued to base our work in the Millennium's Development Objectives (UN), in UNEP's mission, in the orientations of the European Community for the Environment and on the reports of the Environmental Panel on Climate Change. A reference to the work of the commissioner for the Environment, Stavros Dimas, whose words, ideas and actions have constituted a source of knowledge, inspiration and motivation.

Because we are teachers, we believe in the human capacity for adaptation, compromise and problem solving. Therefore, we think that only the informed and formed human beings, in a global sense, may be the vehicle for the resolution of the environmental, political, economic and humanitarian planetary problems.

We are conscientious that the targets of education have changed. However the Curricula and informal education has not yet catch up with the historic development of societies, economies and values. The set of principles, society, economic prosperity, environmental, political paradigms, are to be reformulated.

The basis of this congress is the *modus operandi* of Basic school teachers, in other words, we understand that knowledge is not compartmented, but transdisciplinary and therefore less reductive in its approaches. It is a paradox, but, the passage to college and specialisation shortens the "window to the world" and the possibility of a global and interdependent vision. We know for a fact that this narrowing is more than needed we just want all to contribute at the same level. No subject is deified nor consider of lesser importance. The Science of Collaboration has sprouted.

Therefore, we decided to unite in this congress, specialists in the areas of:

• Climate Change, or Climate Disaster as some begin to call it;
• Preservation of biodiversity;
• Changes in the Ocean's biodiversity;
• Carbon cycle and the role of Oceans;
• Use of fresh water and ocean's by mankind;
• Humanity's risk of survival;
• Mentality and behaviour changes;
• Violence in its juvenile, ethical, religious and geographical aspects;
• Extreme poverty;
• Environmental education at school, home and society;
• Ethics;
• Environmental History.

We think that by this approach we will contribute to the boosting of knowledge, skills and experience exchange platforms. The decisive change in attitudes, by means of real capture and retaining of ethical and scientific principles and of shared responsibility, seems to us a gateway towards a meaningful mind-set and well-being.

The events that triggered this book were a gathering of the greatest minds working on the subjects and themes from all over the world, and educators (teachers, parents, general public) that were willing to learn more, to share and enhance their work. We have done our very best so these meetings would allow them to pass this knowledge to a younger audience in a correct, factual way and help them design educational projects or change their everyday life behaviour in a skilled and efficient manner. It was also aimed at scholars, investigators, professionals and entrepreneurs who wished to deepen their knowledge of these subjects.

After the II ISC we maintained correspondence with many speakers who mentioned many times that we were world level pioneers in this type of actions.

Convinced that, independently of our 'smallness', our work had value, we were motivated to find simple and effective Environmental Education tools, without using scare-tactics.

For all that matters, teaching through example is action. Hence, being educators we hope to be the best example possible.

Therefore, the battle is for a secular humanistic education that can construct structured, loved and efficient citizens, armed with the ethical tools that will allow them to provide, to the coming generations they will educate, humanistic and holistic principles. Some of these principles are for them to figure out.

Our objectives were:

- Contribute to solving pressing social, cultural, environmental and educative matters;
- Inform about some realities that have planetary level impact;
- Contribute to the management of natural resources and the preservation of cultural heritage;
- Understand that there does not have to be a choice between a sustainable environment and a strong economy;
- To raise awareness of teachers in schools and universities, parents and general population to their responsibilities as active agents affecting the education of environment concerns in their communities, as well as agents of political change;
- Present, share and discuss adequate and innovative educational experiences;
- Modernising the educational intervention themes in accordance with the global reality;
- Adequate the educational practises, making them in line with the holistic education of people, emphasising high level mental operations;
- Promote humanistic values and scientific rigour when solving problems;
- Unite, in the same space, the central and local power, educators, entrepreneurs, artists, researchers and NGO's;
- Create work and cooperation platforms between schools, communities, and central and local power;

- Create work and cooperating platforms between schools and universities;
- Benefit from sharing geographical and culturally different pedagogical experiences;
- Contribute to the implementation of environmental themes in the Basic and in service training for teachers;
- Instigate creativity as a pedagogical tool;
- Promote the notion of community and community action;
- Promote self-agency and leadership strategies;
- Contribute to the development and interest of Science, Arts and new Professional options.

The II ISC aimed, therefore, to be an encounter of citizens of the world, focused in the cooperation and resolution of problems that affect us all, creatures of Nature.

This outcome, this book, its chapters were chosen in order to give an overview of what we are doing.

Educators, parents and students, are all invited to go through the chapters and let the sparkle of curiosity take them further.

All of us awake and aware that we are all needed, that LIFE depends on our common responsibility and capacity to create solutions.

We shall carry on 2012 in Beira/Gorongosa National Park, Mozambique. Our goals will be somehow different, our philosophy the same: to summon the best, to call upon all, regardless their age, their literacy level. We must see us as a learner's community, fearless to absorb all we are able to and ready to share what we "have got". This Congress should be an exploratory meeting of minds, situations, knowledge, cases and whatsoever you want to call it. "The making of...", this work in progress trade mark we have, should, as never, be our major goal as well as our tool.

Everyone is invited to replicate our work.

Tip: "Knit" for your life and for your Planet!
It does not have to be a green sweater, as in jade and trees, it may be orange, like dusk and dawn, blue, like sky and sea, grey like rocks and birds, yellow like sun and day, red like tongues and flowers, black, like oil and water snakes, purple like veins and hats, colourless like mind and pain.

P. S. This is from Michael Marzolla.
The moment Michael was arriving to Braga, he apologised.
As the first thing he saw was a huge, mega outdoor, from McDonald's.

Acknowledgements

This book was the natural follow up of the 2nd International School Congress: Natural Resources, Sustainability and Humanity, held in Braga, Portugal from 5 to 9 May 2010, and felt like a collaborative exercise.

We are deeply indebted to our major sponsors: the Ministry of Education, particularly Prof. Drª. Maria de Lurdes Rodrigues, the former Minister of Education, and also to the Direcção Regional de Educação do Norte (Northern Regional Board of Education), mainly its Director, Dr. António Leite.

It was our privilege to work with several Institutions and Association: The Secretariat of the Convention on Biological Diversity, UN; Rachel Carson Center, Munich; Museu D. Diogo de Sousa, Braga, Pt; Gorongosa Restoration Project, Mozambique; Universidade do Minho, Braga, Pt; European Society for Environmental History; National Commission for UNESCO; European Commission; National Geographic—Portugal; Association of South Asian Environmental Historians; Science in School; Forest History Society, USA; American Society for Environmental History; Portuguese History Academy; National Refugee Committee; UNICEF; NATO; International Amnesty; Braga City Hall; Casa do Professor; Arte na Leira; Ourivesaria Freitas; Portuguese Youth Institute; Xunta de Galicia; Caravela Travel Agency; Pousadas de Portugal; Peneda-Gerês National Park; Rádio Universitária do Minho; Arboreto Barcelos; Centro Interpretativo de Carvalho de Calvos; Parque Natural do Litoral Norte; Luís Coquenão; Sofia Brás Monteiro; Ordem dos Biologos; Official Journal of the Academic Association of the University of Minho; Traditional Goldsmithery Museum; SketchPixel and LKComunicação, and all schools engaged in this project.

Our heartfelt thanks to all great speakers from all over the world, who kindly shared with us their knowledge and researches. We owe a great many thanks for their notable contributions.

We also thank the directors of the three schools involved in the organization of the Congress, and appreciate the work done by their teachers and students.

Our special thanks to Stella Mendonca, soprano, accompanied by Paul Suits, piano. This gesture of solidarity has drawn attention to the children of Mozambique to our common project *Building a school, building a future.*

Last but not least, the president's special thanks to all volunteers. Maria José Marques and Nuno Castro and Marta Mendonca, words cannot describe your good will.

Contents

Contributors

Pedro M. Antunes Research Chair in Invasive Species Biology Department of Biology, Algoma University, Sault Ste. Marie, Ontario, Canada

Mateusz Banski Secretariat of the Convention on Biological Diversity, Montreal Canada

Laura Barraza Faculty of Arts and Education, Deakin University, Melbourne, Australia

Fernando JAS Barriga Department of Geology, University of Lisbon, CREMINER-FCUL and LARSyS Associated Laboratory, Lisbon, Portugal

Barbara Bodenhorn University of Cambridge, Cambridge, U.K.

Alexandra Ferreira de Carvalho Departamento de Prospectiva e Planeamento e Relações Internacionais, Ministério do Ambiente e do Ordenamento do Território, Lisboa, Portugal

Ranjan Chakrabarti Jadavpur University, Kolkata, India

Filipe Oliveira Costa Departamento de Biologia, Universidade do Minho, CBMA – Centro de Biologia Molecular e Ambiental, Braga, Portugal

S. L. Dashiell Bren School of Environmental Science & Management, University of California, Santa Barbara, USA

Ahmed Djoghlaf Secretariat of the Convention on Biological Diversity, Montreal, Canada

Joan Freeman Middlesex University, London, UK

Mário Freitas Santa Catarina State University (UDESC), Geography Department, Environmental Studies Centre (NEA), Joinville, Brasil

H. Grossman Department of Education, University of California, Santa Barbara, USA

Zenita C Guenther Centro para o Desenvolvimento do Potencial e Talento, UFLA, Lavras, Brasil

J. Donald Hughes Department of History, University of Denver, NJ, USA

Anthony Michael Marzolla Cooperative Extension 4-H Youth Development Program, University of California, Santa Barbara, USA

Telmo Morato Department of Oceanography and Fisheries, University of Azores IMAR–Institute of Marine Research and LARSyS Associated Laboratory, Horta (Azores), Portugal

Dora Aguin-Pombo Universidade da Madeira, Madeira, Portugal.
CIBIO, Centro de Investigação em Biodiversidade e Recursos Genéticos, Universidade do Porto, Vairão, Portugal

Ricardo Serrão Santos Department of Oceanography and Fisheries, University of Azores IMAR–Institute of Marine Research and LARSyS Associated Laboratory, Horta (Azores), Portugal

A. Yau Bren School of Environmental Science & Management, University of California, Santa Barbara, USA

Chapter 1
What Does Environmental History Teach?

J. Donald Hughes

Abstract Environmental history studies the mutual relationships of humans and nature through time. Historians and others are active in this field in many parts of the world, the literature is vast and growing, and the subject is taught in schools and universities. Its audiences include students, other scholars, policy makers, and a general public, all interested in environmental issues of great import in the modern world. But what does environmental history have to say to these audiences? What are its lessons?

First, it teaches that human history cannot be understood apart from nature. The environment is not just a backdrop for the stage of human politics, wars, and culture; it is a series of influences that interact with every human activity. Environmental processes are important in human history, and it is just as important to take account of human influences on ecosystems and natural areas.

Second, it teaches the importance of science to historians in tracing the interaction of humans and nature. Historians can only rarely be scientists, but they must be familiar with what science says about their fields of concern. Traditional historical sources must be supplemented by studies of changes in climate, ecosystems, and resources. Examples of the integration of scientific and historical evidence are adduced. The second lesson has to do with method, and lies along the continuum between history and science.

Third, it teaches that present-day environmental issues and concerns have their roots in the past, and that research to understand their precedents is valid and rewarding. The study of past effects of environmental forces on human societies, the impact of human activities on the environment, and the development of environmental attitudes and understanding, gives needed perspective to the dilemmas of the contemporary world. This dimension reveals continuity between the past and the present insofar as human-environmental relations are concerned.

Fourth, it teaches a perspective of scale. Local changes inevitably occur within the processes of the planetary environment. The oceans, the atmosphere, the magnetosphere, and cycles of elements are worldwide phenomena, and they affect events in every region and locality. Their effects may be shown most instructively in more

J. D. Hughes (✉)
Department of History, University of Denver, 25 Spring Street,
Apt. No. 302, 08542 Princeton, NJ, USA
e-mail: dhughes@du.edu

A. Mendonca et al. (eds.), *Natural Resources, Sustainability and Humanity,*
DOI 10.1007/978-94-007-1321-5_1, © Springer Science+Business Media Dordrecht 2012

limited case studies, but no case study, however small, may be considered in isolation. This dimension is one of scale and considers time and space not as opposites,
but as coordinates of definition.

These are four lessons of environmental history; it is not suggested that they
are the only ones. There are others equally far-reaching, as well as more specific
lessons derived from particular studies. Most importantly, environmental history
offers methods and perspectives that are crucial to the decisions now being made
as human society faces choices about our response to global environmental crises.
We must learn the lessons of environmental history in order to make wise decisions
in the present.

1.1   Introduction

Environmental history studies the mutual relationships of humans and nature
through time. One of its leading exponents, Donald Worster of the University of
Kansas, described the field of environmental history as consisting of three levels.
The first deals with nature itself as it affects human history. The second shows how
the human socioeconomic organization interacts with the environment, causing re-
ciprocal changes. The third is intellectual, showing how individuals and groups describe and regard nature in the realms of literature, philosophy, religion, and popular
culture (Worster 1988a, pp. 289–307, discussion on p. 293).

Historians and others are active in this field in many parts of the world, the literature is vast and growing, and the subject is taught in schools and universities. Its
audiences include students, other scholars, policy makers, and a general public, all
interested in environmental issues of great import in the modern world. But what
does environmental history have to say to these audiences? What are its lessons?
William Cronon of the University of Wisconsin discussed these questions in an essay entitled "The Uses of Environmental History," In which he made offered four
major answers. First, all human history has a natural context. Second, change is
unavoidable, that is, neither nature nor culture is static. Third, all environmental
knowledge, including our own, is culturally constructed and historically contingent.
Fourth, environmental historians offer understanding of the past, not prescriptions
for the future (Cronon 1993).

In my own attempt to answer the question, "What does environmental history
teach?" I will borrow some ideas from both Worster and Cronon, but I will make
a somewhat different emphasis, and will use visual images from different parts of
the world to illustrate my points (Hughes 2008). What does environmental history
teach? First, it teaches that human history cannot be understood apart from nature.
Second, it teaches the importance of science to historians in tracing the interaction
of humans and nature. Third, it reveals continuity between the past and the present
insofar as human-environmental relations are concerned, so that the past provides
perspective, and present issues find their roots in the past. Fourth, it teaches a balance of scale, because local changes inevitably occur within the processes of the

planetary environment. In other words, the local and global always affect one another.

1.2 History is not Apart from Nature

First, then, environmental history teaches that human history cannot be understood apart from nature. The environment is not just a backdrop for the stage of human politics, wars, and culture; it is a series of influences that interact with every human activity. Environmental processes are important in human history, and it is just as important to take account of human influences on ecosystems and natural areas.

Here is an image to illustrate that idea (Fig. 1.1). When, flying above the Great Plains of the United States, one has a window seat and looks down at the landscape, one sees a remarkably uniform pattern of squares, half squares and quarter squares formed by roads, fields, and subdivisions.[1] These represent the townships, 6 miles to a side, the ranges, the sections 1 mile^2, and the quarter sections each containing 160 acres, set out by the Federal Land Survey in what were then public lands, beginning in 1785. This was also the framework of the Homestead Act of 1862, which provided for the transfer of public land into the hands of citizens who settled on the land and cultivated it. The pattern displays the application of theory to the natural environment. The theory in this instance happened to be that the proper relationship of a citizen to the land was to own it and to cultivate it, and furthermore that since all citizens in the republic were equal, the land allotted to each was equal in size. I mention it because it is a spectacularly visible case of the effect of human policy on nature.

It is not the only such example. Parts of Italy bear to this day the pattern of Roman centuriation, begun by generals who rewarded their faithful surviving legionaries with gifts of land: a hundred *jugera* to each, a *jugerum* being the amount of land a farmer could plow in 1 day with a team of oxen. Ancient China had a traditional method of land distribution called the well-field system, which divided a square of land into nine smaller equal squares, each of the eight outer plots being assigned to one farm family, with the center plot being a public field cultivated by all eight families with the produce going for taxes.

Nature does not always cooperate with human policy, of course, nor does it always passively accept the organization that is forced upon it. Arbitrary squares take no account of such fundamental features as springs, streams, and variations in productivity and exposure. As the frontier moved west across the nineteenth-century territory of the United States, it gradually became apparent that 160 acres might be adequate for a farm in the tall-grass prairies with their relatively generous rainfall, but the same area could spell crop failure and starvation on the short-grass plains where, unfortunately, the proverb that rain follows the plow proved to be untrue. We must take account of the culture-nature continuum.

---

[1] See Hughes (2006, p. 43, illustration 7).

Fig. 1.1 Great Plains of North America

Environmental history includes nature and culture concurrently. In its simplest terms, it requires that a study can be useful only if it considers and correlates change both in human societies and in the aspects of the natural world with which they are in contact. The relationship between the two is in almost every case that of recipro-cal influence. A change made by humans in the environment virtually always re-dounds and generates change in cultural conditions. A history that does not include both terms cannot be called environmental history in the sense intended here. This assertion may seem self-evident to many of those who work in environmental his-tory as a subfield of the historical discipline, as well as to many historical geogra-phers, although there are a few who contend otherwise.

In the late twentieth century certain historians turned to the hitherto obscured ac-counts of those who had seemingly lacked power: to women's history; the histories of racial, religious, and sexual minorities; even the history of childhood. It is an understandably tempting extrapolation to look at environmental history as part of this progression. In the pyramid of power, the beasts and trees, and Earth herself, occupy the lowest stone courses that support the structure. Historians can now dem-onstrate that these supposedly voiceless and largely defenseless entities were in fact authentic actors in the historical drama and include them, too, in the larger narrative. As ethical extension has granted roles to immigrants, women, and former slaves, and recently has considered whether trees should have rights, so a similar historical extension can now grant narrative attention to other living things and the elements[2].

---

[2] This idea is found in Nash (1985).

To see environmental history simply as part of a progression within the discipline of history would, however, be a serious mistake. Nature is not powerless; it is, properly considered, the source of all power. Environmental history is useful because it can add grounding and perspective to the more traditional concerns of historians: war, diplomacy, politics, law, economics, and technology. It is also useful because it can reveal relationships between these concerns and the underlying processes of the physical world. Nature does not meekly fit into the human economy; nature is the economy that envelops all human efforts and without which human efforts are impotent. History that fails to take the natural environment into account is partial and incomplete.

To illustrate the principle that history cannot be fully understood apart from nature, I would like to present an image of an Egyptian rural landscape at the margin where the cultivated land borders the desert. A section of the vast Sahara occupies the upper portion of the view, stretching into the distance. Just below it is a white-painted village recently built of clay bricks. On the near side of that are irrigated fields planted with cotton, wheat, and other crops, and toward the bottom of the scene, a widely spaced grove of date palms stands among the fields. What we see in this picture is the meeting place between what the ancient Egyptians called the "Red Land," the dry and interminable realm of Set, the god of windstorms, and the "Black Land," the fertile watered soil favored by Osiris, god of plants and cultivation. This is a landscape that can be explicated by environmental history, but only if both natural history and cultural history are included as terms in the definition.

I have colleagues who maintain that environmental history is simply the history of the environment. This is because they define the environment as including "climate, geology, and geomorphology, not living things" (Grove and Rackham 2001). A history that included living things, they insist, should be called ecological history. Even with the change in terminology, however, they would focus attention on changes in the landscape, not social, economic, or other cultural changes. In the Egyptian image before us, therefore, they would focus their investigation on the desert, noting the long geological and climatic record of a terrain that was formerly provided with plentiful rainfall and subjected to water erosion. Climatic transformations associated with the end of the most recent ice age shifted the wind patterns, and by the period around 5,000 years before the present achieved a physical regime not dissimilar to that of the present. There is no doubt that these observations are useful, nor any doubt that an environmental historian needs to know them to help in the reconstruction of the past, but the environmental historian must always keep human history and anthropogenic changes in the center of the narrative.

Let us look, however, at the other end of the spectrum for a moment. It would certainly be possible to see the scene before us as an illustration of modern Egyptian political-economic history, as part of the events following the Revolution of 1952. Gamal Abdel Nasser had become president of Egypt and had determined to make the new Arab Republic an industrialized, secular, self-sufficient society that could hold its own in the global market economy. To do this, he considered the construction of the high dam at Aswan one of the first priorities, in order to generate electricity for industrialization, prevent disastrous atypical periodic floods, and to provide

Fig. 1.2 The Nile Valley of Africa

dependable water for irrigation that would not only allow year-round multicropping of food and exportable cash crops, but would also make possible the extension of productive cultivation into formerly desert areas such as the one in our image. All of this is true, and familiarity with it is a necessary prerequisite for the work of the environmental historian, but it does not provide us with an environmental-historical narrative.

It is the work of the environmental historian to master both of the preceding accounts, to relate and combine them, but more importantly to see what each of them has left out, particularly in terms of causal interrelationships that may have escaped them. In the space available, I cannot attempt to give a full analysis of this scene, but I can briefly ask a few questions to indicate the kinds of issues that environmental historians might investigate here. Why does the village occupy a belt of land between the sown land and the desert? Is it the desire of the peasants not to allow structures to intrude on the productive soil? But where did the clay to make the bricks in the houses come from? One might observe that it traditionally came from the silt deposited by the River Nile, and that the Aswan High Dam now prevents that silt from reaching Lower Egypt. How rich is this newly irrigated desert soil, and does it require a major application of industrial fertilizer to keep it in production? What is the state of the soil ecosystem and the various forms of life that make it up? How much of the land produces food for the rapidly growing Egyptian population, and how much bears cotton and other export crops? Again, after the building of the High Dam Egypt became a net importer of food. What is the rate of agriculturally related diseases as schistosomiasis? Is this land subject to salinization, as is so much of the irrigated land in Egypt?

It is not my intention to provide a case study, but simply to indicate the importance of an investigation of questions including both culture and nature in any environmental-historical project. Some might argue that to frame the inquiry in these terms makes environmental history an anthropocentric enterprise. So it is, but it must never lose sight of the ecological impacts and the costs to other forms of life and to the environment itself (Fig. 1.2).

1.3   The Importance of Science to History

Second, it teaches the importance of science to historians in tracing the interaction of humans and nature. Historians can only rarely be scientists, but they must be familiar with what science says about their fields of concern. Traditional historical sources must be supplemented by studies of changes in climate, ecosystems, and resources. In fact, environmental historians use the tools of both history and science, and thus attempt to bridge the gap between what C. P. Snow called *The Two Cultures* within the modern academic community (Snow 1959). On the one hand, environmental historians, being historians, must be consistent and thorough in their employment of the historical method, searching out all the available written sources, subjecting them to external and internal criticism, and interpreting them carefully. But in order to understand the environment, they must become fluent in the language of natural science and be able to use what science can tell us about the realms of history that we choose to study. As Snow said, the failure to comprehend both sides of the cultural divide "is leading us to interpret the past wrongly, to misjudge the present, and to deny our hopes of the future" (Snow 1959, p. 60).

As an example of the need for environmental historians to use the scientific method along with the historical method, I offer the image of Easter Island, named Rapa Nui by the modern Polynesians, with its huge anthropomorphic statues of volcanic stone called *moai* standing in a virtually treeless landscape (Fig. 1.3). Specifically, it is Ahu Tongariki, where 15 moai form a row, facing away from the sea.[3] The case of Easter Island has become a textbook example of a society that destroyed its own resource base through deforestation and overpopulation and duly suffered a collapse ending in a small population and much of the island in ruins. The story has been included in global environmental histories such as Clive Ponting's *Green History of the World* and in Jared Diamond's *Collapse*, where it is told more effectively (Ponting 1992; Diamond 2005). How can an environmental historian arrive at a viable explanation that corresponds relatively truthfully to what happened on Easter Island before and after the time of European discovery?

---

[3] See illustration 14 in Hughes (2006, p. 82).

Fig. 1.3 Easter Island in the Pacific Ocean west of South America

Written historical sources take us only so far. There are the accounts of the discoverers, including ships' logs. On 5 April 1722, Dutch commander Jacob Rogeveen sighted and named the island, describing it as largely bare of trees, with a small population and with a number of the large statues standing. Other explorers, including the English Captain James Cook, followed. There are writings of nineteenth-century missionaries and twentieth-century anthropologists. These reveal that all the statues remaining erect were pushed down by the islanders by the mid-nineteenth century. The ones now standing were re-erected by Europeans, Americans, and Japanese in the twentieth century. When we turn to the crucial pre-European period, traditional historical method gives us very little. There was a native written language on Easter Island, but when most of the population was enslaved and taken off island in the mid-nineteenth century, the knowledge necessary to read it was lost, and it remains undeciphered today. Oral traditions offered interesting but fragmentary hints. From these sources alone, the ecological disaster is inexplicable, although they contribute to an historical account.

The sciences provide much of the needed evidence for that account. The archaeologist is here, as for many of the efforts of ancient historians, a valuable co-worker. Radiocarbon dating suggests that the first human occupation of the island began between 600 and 800 AD. In several parts of the island, trenches reveal a layer of closely packed root casts of palm trees similar to the Chilean wine-palm *Jubaea chilensis*. Hoards of tiny coconut-like fruits of such palms are found in caves. There is evidence of many other species of trees. Pollen analysis reveals the presence of forests until about 500 years before the present. All of this indicates that a relatively densely forested island became deforested during the period of human occupation. Dwelling sites show agricultural activity almost everywhere on the island by 1500. More recent structures include stone-lined planting pits and lithic mulch designed to protect plants from the wind, which would have affected them more after tree removal. It is evident that in the environmental history of Easter Island, historical source work and science supplement one another marvelously.

Scientific techniques studying ecosystems, biodiversity, climate, introductions of organisms, diseases, atmospheric chemistry, and many other factors of change, are of obvious use to the environmental historian no matter what chronological period or geographic regions constitutes the chosen area of study. This is particularly true of ecological science. Environmental history developed in some measure out of the recognition that ecological science has implications for the understanding of the history of the human species. One of the implications is that human civilizations, even those of advanced technological cultures, cannot place themselves outside the principles of nature. Ecology places the human species inside the web of life, dependent on it for subsistence and survival. One cannot deny the importance of scientific literacy in principle for environmental history, quite apart from the obvious practical difficulties this presents for the preparation and continuing education of environmental historians.

As John McNeill said, "The enormity of ecological change [today] strongly suggests that history and ecology, at least in modern times, must take one another properly into account. Modern history written as if life-support systems of the planet were stable, present only in the background of human affairs, is not only incomplete but is misleading. Ecology that neglects the complexity of social forces and dynamics of historical change is equally limited. Both history and ecology are, as fields of knowledge go, supremely integrative. They merely need to integrate with one another" (McNeill).

1.4 Environmental Concerns are Rooted in the Past

It may seem that awareness of environmental issues arose only in recent times, as the result of concern about problems such as pollution and the scarcity of natural resources. Environmental history, however, has discovered that there were antecedents to these modern perceptions.

A criticism of environmental history sometimes raised by other historians is that of presentism. These critics note that awareness of environmental problems is entirely a contemporary phenomenon. The very word 'environmentalism' did not emerge in general use until the 1960s, and environmental history became a recognizable subdiscipline only in the 1970s. The motive that led to the inquiry was a reaction to uniquely modern problems. Is environmental history, therefore, an untenable attempt to read late twentieth century developments and concerns back into past historical periods in which they were not operative, and certainly not conscious to human participants during those times? The problem with this criticism is that it is fundamentally an argument against history itself as an intellectual endeavor that can be applied to the understanding of the present. Modern problems exist in their present forms because they are the results of historical processes. The relationship with nature was the earliest challenge facing humankind. It would take a particularly egregious form of denial not to see a precedent for the market economy in the exchange of a tribal nomad's meat and skins for a village agriculturalist's grain and textiles. The Greek philosopher Plato described soil erosion, and the Roman poet Horace complained about urban air pollution (Hughes 1994). The Columbian transfer of Europeans along with their crops, weeds, animals, and diseases to the New World in large part explains the history and present state of the Americas (Hughes 1994). The study of past effects of environmental forces on human societies, and the impact of human activities on the environment, gives needed perspective to the dilemmas of the contemporary world.[4]

An image that illustrates this relevance of the past is a view of the eroded slopes of Mount Parnes in Attica, not far from Athens. Of all the passages from ancient writers on the natural world, the most fascinating for modern environmental historians may be Plato's comment on the consequences of deforestation in just such a place: "What now remains compared with what then existed is like the skeleton of a sick man, all the fat and soft earth having wasted away, and only the bare framework of the land being left" (Plato). Mediterranean environmental history is in large part a history of deforestation and its consequences, although much more than that. Recent work in fields such as anthracology, palynology, and computer modeling of climate change provides new evidence supporting the judgment that ancient deforestation represents environmental damage that contributed to disruption of Mediterranean economies (Hughes 2010). Of course other factors were operative, notably agricultural decline.

The passage by Plato in the *Critias* just quoted is valuable ancient evidence for ancient deforestation, but it has received seemingly endless commentary, some of it consisting of quibbles intended to explain it away. Although it occurs in a section about the lost continent of Atlantis, it is not at all mythical. Plato describes evidence present in his own time and place, and invites his audience to look at it: timber in standing buildings that had been taken from mountains left only with "food for bees" (flowering annuals and shrubs), and shrines marking springs that had dried up when the forests were removed (Fig. 1.4). Elsewhere (*Laws*) Plato advises planting trees to improve water supply.

If the past is not usable, then history is an enterprise in vain. The Greeks did not have science in the modern sense, since dependence on method including observation and experiment was only partially achieved. But it would be gratuitous to ignore the steps they took in the direction of science. In particular Theophrastus was asking many questions that modern ecologists were to ask about the relationships of organisms to their environments, to one another, and to humans. He rejected the

---

[4] See Worster (1988a, pp. 289–308; 1988b, pp. 3–22).

Fig. 1.4 Mount Parnes in Attica, Mediterranean Europe

excessive teleology of Aristotle, but Aristotle also made some interesting assertions about animals that seem, in retrospect, ecological. For example, Aristotle's observation that animals seek out surroundings that are "appropriate" (*oikeios* in Greek) may, through the word coinage of Ernst Haeckel, be the root of the modern word, "ecology." From earliest times human societies have lived in interaction with, and in dependence on, the natural world that surrounds them and, indeed, includes them. They had a dawning awareness of their situation, and expressed it in various ways. All this is a valid and rewarding subject for historical study.

1.5   Scale in Time and Space: Global and Local

Third, environmental history teaches a perspective of scale. Local changes inevitably occur within the processes of the planetary environment. The oceans, the atmosphere, the magnetosphere, and cycles of elements are worldwide phenomena, and they affect events in every region and locality. Their effects may be shown most instructively in more limited case studies, but no case study, however small, may be considered in isolation. Time and space are not opposites, but coordinates of definition. What I have in mind is that environmental history in its essence and by definition implies a vast sweeping perspective, inclusive of the environment by being global and inclusive of history by stretching from origins to the present and even, dangerously, peering into the mist-obscured future.

First, let us look at time. My contention is that the field of environmental history investigates every time period in human history, including prehistory, ancient, medieval, and modern. Individual studies may take shorter periods as their frames of analysis, but the scope of the enterprise of environmental history is only limited by the consideration that human societies were interacting with the natural environment, not that there was any particular mode of interaction, or any particular form of recognition of the nature or extent of the interaction at the time. In particular I reject the common, if unarticulated, idea that environmental history should be exclusively, or almost exclusively concerned with the modern world because of the rate of change and environmental awareness that exist in recent times. The ancient and medieval periods, in which perhaps the majority of human modes of environmental relationships and the institutions that enact them originated and developed toward their modern expressions, also merit careful study in environmental history.

The image I use here is one section of a vast system of ancient agricultural terraces in Jordan near the ancient Nabataean city of Petra (Fig. 1.5). They are now almost completely abandoned, although there is a small settlement in view, a few trees and gardens, and here and there people and domestic animals can be discerned in what otherwise seems overwhelmingly an arid and empty scene. These terraces were constructed in a time of greater prosperity to prevent erosion and to allow the planting of crops on a steep hillside. Here, written in the landscape itself, is the story of the rise and fall of human civilization in one of the lands of its early birth, and it is a rise and fall mediated by the interaction of humans and the environment. Here the story of the uneven path of civilization due to environmental problems such as deforestation, erosion, and agricultural exhaustion can be traced. Within this single view there are several miles of terrace walls painstakingly built of heavy stone. The labor expended in their construction was considered worthwhile, justified economically and in terms of human calories because they made food production possible by preventing erosion on the steep gradient. At that time, forests on the higher elevations evened out the downhill flow from seasonal rainfall and provided a dependable water supply from springs and local streams that then existed. But the incessant cutting of trees for construction and fuel in the nearby city, and the grazing of goats and sheep on the denuded heights, assured the permanent destruction of forests and the water supply that depended on them, so the hills are now dry and the terraces bear almost nothing. Such an image is both an important phase of environmental history and a cautionary tale that warns historians to widen their awareness of the deeper past.

What holds true of time for environmental history also holds true of space. That is, however we may choose to circumscribe the areas for particular studies, the whole Earth is our subject. Perhaps it extends even beyond that, since energy from the Sun and tides caused by the Moon are also important environmental influences. Just as every modern historical moment is connected to a long formative past, so every locality or region exists within the setting of the ecosphere, and historians neglect that fact at their peril. Even to write the environmental history of a single

Fig. 1.5 The arid land of Jordan in western Asia

garden requires a sense of its place on the planet. Practically speaking, each study must be grounded in a delimited space and in a certain period of time, because research and writing must have a stop, at least until the next book. But serious environmental history by its very nature must recognize the many links to a larger and inclusive system.

1.6   Conclusion

The final image I present is that of the modern Bay of Naples (Fig. 1.6). Naval vessels head out of the port. A medieval fort occupies a point of land. The suburbs climb up the slopes of Vesuvius. Applying the three lessons I have presented, what can be learned in regard to the environmental history of this place?

First, looking at culture and nature, it is clear that the interplay of land and sea have helped to form the city as an active and economically viable center, here on this bay. It has been a naval headquarters at least since Roman times. The local culture has adapted to changes in sea level; there are Roman temples here whose columns stand in seawater; obviously they were on dry land when they were built, but water lines and seashells show that the water was occasionally even higher than it is now. This is due to local changes in elevation produced by an active geology, bringing up the second point.

Fig. 1.6 Bay of Naples in Italy, Mediterranean Europe

Second, looking at history and science, it is well known that Vesuvius erupted in 79 AD and destroyed several cities along the shore of the Bay of Naples, including Pompeii and Herculaneum, with great loss of lives. We have, for example, the contemporary account of Pliny the Younger, who witnessed the eruption and described it. Science can testify that there have been many other eruptions and earthquakes here, that the soil in this area is fertile because it is volcanic in origin, and that Vesuvius is still active. It erupted during the Second World War, for example.

Yet people continue to build close beneath it as if there were no danger. There is perhaps a parallel here with the present threat of global warming, bringing up the third point, the continuity between past and present. That is, a concern for environmental security must take account of the volcanic history of this unstable region, and architecture, infrastructure, city planning, and measures for inevitable disaster should all reflect the experience of the past.

Fourth, looking at time and space, it is evident that a complete understanding of this modern scene depends in large part on knowing the ancient and medieval history that produced it. And Naples, connected through its bay with the Mediterranean Sea to the world ocean, and through the atmosphere to everywhere else on Earth, cannot truly be considered in isolation.

The six images we have seen depict five regions of Earth: the Great Plains of North America, the Nile Valley of Africa, Easter Island in the Pacific ocean west of South America, the arid land of Jordan in western Asia, and Mount Parnes and Naples in Mediterranean Europe. They represent samples of the global extent of environmental history.

I have touched on only four of the lessons we can learn from environmental history. There are, of course, many others including some that are limited and specific. Environmental history gives us information about the past that forms an indispensable perspective on the present and future.

References

Cronon W (1993) The uses of environmental history. Environ Hist Rev 17(3):1–22, particularly 12–18

Diamond J (2005) Collapse: how societies choose to fail or succeed. Viking, New York, pp 79–119

Grove AT, Rackham O (2001) The nature of Mediterranean Europe: an ecological history. Yale University Press, New Haven, p 376

Hughes JD (1994) Pan's travail: environmental problems of the ancient Greeks and Romans. Johns Hopkins University Press, Baltimore, pp 73, 149

Hughes JD (2006) What is environmental history? Polity Press, Cambridge, p 43 (illustration 7), 82 (illustration 14)

Hughes JD (2008) Three dimensions of environmental history. Environ Hist 14(3):319–330

Hughes JD (2010) Ancient deforestation revisited. J Hist Biol 43(3):36–51

McNeill JR (2001) Something new under the sun. W. W. Norton and Co., New York, p 362

Nash R (1985) Rounding out the American revolution: ethical extension and the new environmentalism. In: Tobias M (ed) Deep ecology. Avant Books, San Diego

Plato, Critias 111 B-D (1929) Plato, Timaeus, Critias, Cleitophon, Menexenus, Epistles (trans: Bury RG). Harvard University Press, Cambridge, pp 272–275

Ponting C (1992) A green history of the world. St. Martin's Press, New York, pp 1–7

Snow CP (1959) The two cultures and the scientific revolution. Cambridge University Press, Cambridge

Worster D (1988a) Appendix: doing environmental history. In: The ends of the earth: perspectives on modern environmental history. Cambridge University Press, Cambridge, pp 289–308

Worster D (1988b) The vulnerable earth: toward a planetary history. In: The ends of the earth: perspectives on modern environmental history. Cambridge University Press, Cambridge, pp 3–22

Chapter 2
New Orleans: An Environmental History of Disaster

J. Donald Hughes

Abstract If Egypt is the gift of the Nile, similarly New Orleans and all southern Louisiana are the gifts of the Mississippi River. Without human interference, the river would continue to add to its vast, flat delta, flooding and shifting from one channel to another. The wetlands, along with grassy marshes, and the barrier islands formed further out on the edge of the Gulf of Mexico, formed insulation against hurricanes. New Orleans, a city that is largely below sea level, has been hit by a major hurricane every few decades, but the earlier ones tended to do less damage due to the protection they offered. Healthy ecosystems served as natural defenses. They were like speed bumps against storm surges. But much has disappeared and the rest is endangered, and the reason why can be explained by the environmental history of the region. Natural protection has been stripped away in large part by human projects.

Hurricanes, like other natural disasters, present a problem for historians. Are they events that happen to people, making humans helpless victims? To what extent can humans control them, or at least modify them and guard against them? And in what ways are humans responsible for the destruction they and their works suffer from natural disasters because of the choices they make? The damage caused by a hurricane depends not just on the force of the storm, but also on what people have done to the land. That includes city planning and activities that weaken and destroy natural entities that might protect them, such as the wetlands of the Mississippi delta.

This presentation gives a brief historical outline of the projects that have altered the flow of water, and the silt it carries, in southern Louisiana, and the ways in which they established the conditions that made hurricanes Katrina and Rita in 2005 so devastating. To prevent floods caused by the river, levees were constructed along its course, shunting the river's load of fertile soil to the Gulf and effectively stopping renewal of land in the delta. The main flood danger to the center of the city had shifted to another front: the south shore of Lake Pontchartrain, and to storm surges driven into the lake from the Gulf of Mexico. This danger was exacerbated by the construction of canals for oil barges through the wetlands, and in particular

J. D. Hughes (✉)
Department of History, University of Denver,
25 Spring Street, Apt. No. 302, Princeton, 08542
NJ, USA
e-mail: dhughes@du.edu

A. Mendonca et al. (eds.), *Natural Resources, Sustainability and Humanity,*
DOI 10.1007/978-94-007-1321-5_2, © Springer Science+Business Media Dordrecht 2012

by the Mississippi River-Gulf Outlet, a ship canal intended for direct access to the Port of New Orleans.

Although much attention after the 2005 hurricanes has been directed to the inadequacy of the floodwalls in the city, it must not be forgotten that he city is inextricably part of the Gulf Coast wetland ecosystem, and that ecosystem has historically operated to insulate the city against a number of dangers. Some planners advocate a regional system of coastal and wetland restoration. This would involve diversion of some of the flow of the Mississippi so that sediment can build up wetland where it has been lost. It would mean establishing new protected areas and limiting development within them, especially the excavation of canals. Cypress trees would be protected. Barrier islands could be strengthened, extended and vegetated. Such a program would be expensive, although perhaps not compared with the damage that has occurred by the neglect of the ecosystem.

In 2005, the city of New Orleans, Louisiana, was subjected to twin disasters: the advent of two major hurricanes, Katrina and Rita. Almost 5 years later, a third disaster exacerbated the aftermath of the first two when Deepwater Horizon, an offshore oil rig operated by BP, exploded and began perhaps the most voluminous spill of petroleum the Earth's seas had received until that time. Hurricanes and oil spills are not unrelated, and environmental history needs to examine the causes of their impact on the city and its surrounding ecosystems (Fig. 2.1).

New Orleans as of August 2005 was a historic city, a cultural center, the birthplace of jazz music, and one of the world's busiest ports. It attracted 10 million tourists in a year (Daly 2006). It was also a city with most of its surface located below sea level, and in the path of a storm that would bring heavy rains and a surge of water from the sea on the 29th of that month. I saw New Orleans 18 months later, after the disastrous hurricane Katrina had breached the levees and flooded more than four-fifths of the city with toxic, sewage-laden water as deep as 6 m (20 ft) in some places. Much of the debris had been cleared away, but the city was still suffering from its wounds. On the shore of Lake Pontchartrain there was a lighthouse smashed over at an angle, and bare pilings in the water where there had been restaurants and theaters. I went to the Lower Ninth Ward with university professors who have studied the city, to see first hand something of what had happened. There were whole blocks where houses had been swept away by water exploding from a breach in the floodwall. Some houses were left smashed and crushed, and a few turned upside down. When I looked at the ground and the sidewalks, I saw shells from the canal along with knives, forks, scissors, and children's toys. Block after block were empty houses still erect but often windowless and gutted, stained with mold. All were marked next to the front door with a large X and the necessary data: the date inspected, the inspecting agency, and the number of dead found. In most cases, the latter figure was zero, but Katrina and its aftermath had killed about 1,500 people in New Orleans. In spite of all that, there were signs of hope: someone had scratched in the concrete at the base of a re-erected electric pole, "Ninth Ward Lives!" I met people wearing T-shirts with the motto, "Re-New Orleans!".

Fig. 2.1 Hurricane Katrina, satellite view

The driver in the taxi that took me from Baton Rouge to the New Orleans air-port was an African-American resident of New Orleans who had been at work in a K-Mart the day Katrina struck. She had brought her 9-year-old son with her. When the floodwall broke, the water quickly rose to chest height. She put her son on an inflated air mattress and managed to get to a bridge. The other side was not flooded, but that was a predominantly White-inhabited neighborhood and police with guns stopped the people from crossing. Helicopters later evacuated old people and chil-dren, including her son; eventually she made her way out to Baton Rouge and was reunited with him. Like our driver, more than half the population of New Orleans resettled elsewhere, in Baton Rouge, Houston, Atlanta, and smaller numbers across America. Approximately 200,000 of an original population of 440,000 now live in New Orleans, and the African-American proportion has dropped from 67 to 47%. Certainly the poor suffered more than the affluent. It took a week to evacuate all of the 122,000 people who were stranded in the Superdome and the Convention Center.

Ari Kelman, my colleague who wrote a fine book on the earlier history of New Orleans entitled *A River and Its City* (Kelman 2006), said that the idea was to "Move the homeless, the elderly, the impoverished, the unlucky, all those poor souls who couldn't get out of New Orleans in time to avoid Hurricane Katrina; move them into the city's cavernous domed football stadium. Anyone who has seen a disaster movie could have predicted what would happen next: Katrina slammed into the Super-dome, ripped off the roof, and knocked out the power, cutting off the drinking water and the air conditioning. Those trapped inside had to be moved again—to Houston's Astrodome, of course. If it's not too callous to say so, the stadium mishap is an apt metaphor for New Orleans' environmental history. The sodden city has long placed itself in harm's way, relying on uncertain artifice to protect it from predictable disasters" (Kelman 2005).

If Egypt is the gift of the Nile, similarly New Orleans and all southern Louisiana are the gift of the Mississippi River. Each was formed of the debris deposited by a great river from a vast watershed draining part of a continent. The Mississippi, before the dams and diversions, carried water and sand, silt, and mud from 40% of the land area of what became the United States, and a smaller area in Canada, over many thousands of years. Without human interference, the river would continue to add to its vast, flat delta, flooding and shifting from one channel to another. Most of New Orleans today is below sea level, because the alluvial soil compacts as it accumulates, and further shrinks when it is drained. While in Louisiana, I took another field trip into the wetlands and bayous, where forests of huge bald cypresses and tupelos flourished in the fresh water brought down by the river. Those wetlands, along with grassy marshes, and the barrier islands formed further out on the edge of the Gulf of Mexico, are rich in fish, alligators, water birds, and other wildlife. They provide a stopping place for 70% of the migratory birds in the Great Mississippi Flyway. In the past they also formed insulation against hurricanes. New Orleans has been hit by a major hurricane every few decades, but the earlier ones tended to do less damage due to the protection offered by these millions of acres of wetlands, forests, and barrier islands. They were like a series of speed bumps against the storm surges of salt water. It was a case of healthy ecosystems serving as natural defenses. But much of it has disappeared and the rest is endangered, and the reason why can be explained by the environmental history of the region. That protection has been stripped away in large part by human projects, not by natural processes alone.

Hurricanes, like other natural disasters such as floods, tsunamis, and volcanic eruptions, present a problem for historians. Are they events that happen to people without choice, often without much warning, and make humans helpless victims? To what extent can humans control them, or at least modify them and guard against them? And in what ways are humans responsible for the destruction they and their works suffer from natural disasters because of the choices they do make? The damage caused by a hurricane depends not just on the force of the storm, but also on what people have done to the land. That includes city planning and activities that weaken and destroy natural entities that might protect them, such as the wetlands of the Mississippi delta.

For European colonists determined to exploit North America in the early eighteenth century, the delta of the Mississippi River was an ideal location, a portal to the immense interior along a myriad tributaries large and small. It was, however, an environment that did not offer an ideal site for a city among its intermittently flooded and unhealthy swamps. The French explorer and governor of Louisiana, Jean-Baptiste Le Moyne, sieur de Bienville, after making settlements at several other places, discovered a crescent-shaped bend where the river had built a natural levee that seemed to him relatively safe from tidal waves and hurricanes, and in

1718 decided to establish a city there, naming it *La Nouvelle-Orléans*. It became
the capital of the Louisiana colony in 1722, a year which foreshadowed future
events when in September a hurricane struck, blowing most of the houses down.
The site of today's French Quarter was on slightly elevated ground (3.7 m, or 12 ft
above sea level) near the river, and therefore somewhat protected from flooding.
Nonetheless, there were floods; the river always presented a threat and the colonists added an artificial levee (a dike or embankment) 1.2 m (4 ft) high on top of
the natural one formed by the river, beginning a process that would last at least
two centuries of building barriers against river floods and raising them increasingly higher. During the French period and the Spanish period that intervened from
1763 to 1800, the main flood threat was from the Mississippi River, and landowners were required to build levees in the sectors of land that they owned. North of
the new settlement, in an area that was to become part of the modern city, cypress
swamps and grassy marsh stretched to the shores of Lake Pontchartrain. This almost completely flat expanse was crossed from east to west by the Metairie and
Gentilly ridges, ancient natural levees of no great elevation along a former path of
the river. These ridges were separated by Bayou St. John, a sluggish watercourse
debouching into the lake.

New Orleans became part of the United States as a result of the Louisiana Pur-
chase from Napoleonic France in 1803. The city soon began to use boat-mooring
fees to compensate landowners for levee building. The new Territory of Louisiana,
and the State of Louisiana organized in 1812, lacked a central flood authority, and
although levees ran along many kilometers of both banks of the river, the protection they offered was uneven and failed to prevent floods that occurred frequently,
often only a few months to 2 or 3 years apart. Frequently breaches in the levees that
caused the floods were some distance above New Orleans and entered the lower
parts of the city from behind. In 1849 the Sauvé Crevasse flood penetrated New
Orleans, including part of French Quarter, and the river undercut levees, making
extensive repairs necessary (Colten 2005, pp. 26–28, 73–75). The state, and increasingly the federal government, took greater roles in flood protection as the value of
development in New Orleans and the Mississippi delta became apparent. Congress
passed the Swamp Land Acts, which allowed the state to sell federal land for money
to construct levees. Meanwhile, the US Supreme Court had ruled that the commerce
clause of the constitution made the federal government responsible for maintaining
navigation on rivers, and Congress delegated that job to the Army Corps of Engineers, whose role grew to a leading one in later years. The Corps built levees on the
lower river leading to the Gulf, and it dominated the Mississippi River Commission
created by Congress in 1879. After the Great Flood of 1927, the Corps was given
oversight of flood control and navigation works on the entire Mississippi and its
branches, and after Hurricane Betsy in 1965, direction of the hurricane protection
system of all southeast Louisiana. Unfortunately, the projects undertaken for flood
control also had the effect of changing the hydrological regime of the floodplain to
prevent the deposition of new soil which formerly maintained the bottomland of
southern Louisiana. The attempt to control flooding, it is clear, depended almost

entirely on building levees, and as this effort continued, the river was gradually contained between two sets of levees from far above New Orleans to the Gulf of Mexico. By constricting the river, the levees raised its level further above the surrounding land, with the result that floods that did occur had the potential to be higher and more destructive. They also shunted the river's load of fertile soil to the Gulf, effectively stopping the renewal of land in the delta (Colten 2006). In addition, the Atchafalaya River, a lower branch of the Mississippi, was straightened and contained between high levees that convey floodwaters rapidly to the Gulf, where the erosional materials they contain are dissipated (Saikku 2005). Most of the Mississippi's flow, however, was not allowed to enter the Atchafalaya and was directed past *New* Orleans. Several engineers pointed out the advantages of gates and spillways that could give river water and river mud alternate pathways to the Gulf, relieving pressure in the main stem and helping to build up the delta. There were experiments of this kind, but for the most part the policy was "levees only," and as far as the city's defense against the river was concerned, the policy was a success; from the late nineteenth century onward, the river levees usually held. In the Great Flood of 1927, the bowl of the city filled when 355 mm (14 in.) of rain fell and the pumps failed. The Army Corps broke a levee downstream at Caernarvon to relieve pressure on the levees next to the city (Colten 2005, p. 142). This action flooded St. Bernard Parish, but it may have been unnecessary. In 2005, none of the levees along the Mississippi River above or in New Orleans failed. The main flood danger to the center of the city had shifted to another front: the south shore of Lake Pontchartrain (Colten 2005, p. 31).

New Orleans grew rapidly in the nineteenth century as a result of its busy port and the success of steamboats in making swift upriver travel possible. The relatively high ground near the river soon filled up with structures, and the obvious direction to expand was northward into the swamps toward the lake. For that to happen, the landscape would have to change. Largely to be cleared were the majestic bald cypress trees; their wood was durable timber, so resistant to rot that it was called "the wood eternal" (Dennis 1988). Cypress forests were exploited in Louisiana from early in the French period (Colten 2003), but although sizeable forest survived toward the lakefront, it was increasingly depleted and the maximum period of removal was from the 1890s through the 1930s, after which there were few trees left that were accessible and of useful size (Conner et al. 2007). Floods only made felling the trees by boat and floating them out easier, but this was wasteful because they were sawn off 4 or 5 m above root level and the tall stumps remained (Saikku 1996). Fortunately in the late nineteenth century the city and state created two large parks, Audubon Park and City Park, and with plans developed in part by John Charles Olmsted, scenic versions of the original landscapes of the city were preserved (Colten 2005, pp. 73–75). The next step necessary before streets could be laid out and houses built was drainage; the accepted way to do that in the early nineteenth century was to dig canals, but the extremely low gradient toward the lake rendered canals ineffective in getting the water out, and in addition provided a way into the city for water driven by storms from the lake. Steam-driven waterwheels were installed to lift water into the canals, but drainage lowered the water table and the ground level and left the canals, provided with their own levees, higher above the city. Electric pumps and new canals provided more effective water removal in the years before the First World War, when there were seven pumping stations and 112 km of canals and the water table had been lowered by as much as 3 m (Kelman 2006, pp. 154–155). Development proceeded toward the lake, from south to north, and the pumping stations stayed where they were first placed, toward the south ends of the canals, away from the lake, instead of at the lakeshore levees, where they could have pumped water directly into the lake and served as barriers. As a Dutch engineer remarked after Hurricane Katrina had demonstrated the weakness of the canal levees, "Why in the world would you invite the enemy deep inside your own camp?" (Heerden and Bryan 2007). The major construction of lakefront protection levees took place in 1922–1934. In the 1970s, the lakefront levees were raised to 4.25 m (14 ft).

Then in the eastern part of the city the Industrial Canal, opened in 1923, connected the Mississippi River to Lake Pontchartrain. About halfway along that canal, it is joined by two other canals from the east, the Gulf Intracoastal Waterway, a navigable inland channel that parallels the Gulf coast from Florida to Texas, and the Mississippi River-Gulf Outlet (MR-GO), a navigation channel 122 km (76 miles) long completed in 1965 to connect the Port of New Orleans directly to the Gulf of Mexico, shortening the distance that ships would otherwise have to traverse along the curves of the river. These two waterways, with their levees, form a "funnel" leading directly toward their T-shaped junction with the industrial canal, setting a scenario for disaster when hurricanes drove surges from the Gulf.

Never a stranger to hurricanes, New Orleans had discovered that the high winds could also drive waters from Lake Pontchartrain into the city. A hurricane in 1947 tossed waves over the lakeshore levees and caused significant flooding. By the 1950s, hurricanes received female names bestowed in alphabetical order, an idea taken from George R. Stewart's novel, *Storm*, in which meteorologists called a Pacific low-pressure system "Maria." Consequently, New Orleans suffered Hurricane Flossy, which burst gaps in the canal levees in 1956, and Hurricane Hilda in 1964. In the following year, memorable Hurricane Betsy brought winds of 260 kph (160 mph) and breached the Industrial Canal, damaging 7,000 homes and 300 industries. Hurricane Camille also breached the Industrial Canal in 1969. Incidentally, the idea that hurricanes should be given feminine names was recognized as an instance of gender discrimination, and in 1978 the official lists of names began to contain alternately masculine and feminine names. It was the luck of the draw that gave New Orleans two female hurricanes in 2005, Katrina and Rita.

Meanwhile, the oil and gas industry in the Louisiana wetlands was expanding to become the largest source of crude oil and the second largest source of natural gas in the United States. Oil platforms were also constructed to access petroleum reserves under the Gulf of Mexico, and oil tankers required delivery to the port at New Orleans, so canals were built through the wetlands to provide access. Combined with those excavated for operations on land, the total distance of cuts and canals in the wetlands has been estimated at almost 13,000 km (8,000 mi). These canals allowed salt water to flow into the wetlands, killing trees and other vegetation and eroding away the land. The longest navigation canal was MR-GO, mentioned above, leading directly from the Gulf of Mexico to the heart of New Orleans, providing a channel for oil tankers to come into the port, but also providing a potential funnel for hurricane surges. This is exactly what happened during Hurricane Katrina. Some engineers predicted this danger and recommended building gates on MR-GO that could be closed in case of a storm, but this was not done because of cost and other objections, some of them environmental.

In July, 2004, the Federal Emergency Management Agency (FEMA) brought emergency officials from 50 parish, state, federal and volunteer organizations together for a 5-day exercise held at the Louisiana State Emergency Operations Center in Baton Rouge to help officials develop joint response plans for a catastrophic hurricane (FEMA 2004). It involved a simulated, computer-generated event called Hurricane Pam. Hurricane Pam was assumed to be category 3, bringing sustained winds of 200 kph (120 mph) and up to 510 mm (20 in.) of rain in parts of southeast Louisiana and a storm surge that would top levees in the New Orleans area. More than 1 million residents were targeted for evacuation and Hurricane Pam was expected to destroy 600,000 buildings. According to the scenario, only a third of the population were predicted to leave New Orleans before the storm hit. This was recognition of the fact that much of the city's population lived in relative poverty, with approximately 127,000 in households in which no one owned a car. Hurricane Pam turned out to be an almost exact prediction of Hurricane Katrina, the first of the two hurricanes that actually hit New Orleans just over a year later. I mention it because it indicates that no one should have said that Katrina was unpredictable. There was earlier warning, indeed, from scientists, historians, engineers, and a newspaper, the New Orleans *Times-Picayune*, which had carried a series of stories in 2002 on the very subject, also predicting a disaster very much like what actually occurred. In spite of that, the federal administration's unpreparedness is epitomized by the slow response of FEMA when the emergency occurred.

The large and destructive hurricane designated Katrina reached the northern coast of the Gulf of Mexico on August 29, 2005, 4 days after it had crossed the Florida peninsula. When it reached the Mississippi delta in Louisiana, it possessed sustained winds of 204 kph (127 mph). It produced a storm surge approximately 9 m (30 ft) high, with waves up to 16 m (52 ft), which did catastrophic damage when it smashed into coastal areas of Louisiana, Mississippi, and Alabama (Curtis 2007). Considered as a single event, it was one of the deadliest and most costly weather-related disasters up to that time in the history of the United States, causing more than 1,833 deaths and about $ 125 billion in economic impact. New Orleans suffered most of the deaths and a major share of the damages. The storm surge of Hurricane Katrina found weak points in the system of levees set up by engineers to protect the city. A surge into Lake Pontchartrain, pushed by winds from the north as Katrina passed by, entered the drainage canals and breached the floodwalls in several places, flooding the northern and central parts of the city. The surge from the east piled through the "funnel" into the Industrial canal and breached its floodwalls in three major places, pouring violently into East New Orleans, the Lower Ninth Ward, and parts of the city to the west.

Oil- and sewage-laden water flooded more than 80% of the city as power failed and the pumps ceased to function. In some places the water level rose more than 6 m (20 ft) above the ground. People who were still in the city were trapped in their attics as water came up, unless they could break through and seek rescue on the roofs, rescue that was all too often very slow to arrive by boat or helicopter, if it arrived at all. Indeed, the slowness of government agencies to respond to the disaster, on the federal, state, and local levels, constitutes a lasting disgrace associated with the Katrina disaster. As Kelman expressed it, "Nor do I suggest that this was a natural disaster. It was not. It was an outgrowth of a host of bad decisions, large and small, of miscalculations, of ignorance, even of hubris. It was, in sum, a byproduct of the city's and the nation's environmental history" (Kelman 2006, p. xviii).

Past efforts to protect New Orleans were in the main attempts to control nature, and although nature time and time again has demonstrated its forces to be uncontrollable, those efforts are parts of plans for the future. New Orleans will not be abandoned any time soon, although if a radical change in the way its problems have been addressed does not take place, that might become an unavoidable alternative. But if not abandoned, it must be defended by working technology. Other hurricanes, some of them more powerful than Katrina, will inevitably hit the city sooner or later, and the danger is exacerbated by expected incremental rise in sea level caused by global increases in average temperature. Projects have been advocated, some of them already under way, to repair the breaches, raise and strengthen the levees and build new ones regionally, put pumps where the canals meet the lake, and complete a lock system on the Industrial Canal. The Army Corps of Engineers has proposed to close the Mississippi River-Gulf Outlet. Perhaps using aspects of the Dutch experience with dikes and gates, at the entrance to Lake Pontchartrain and elsewhere, will help. Projects like these are necessary, but they will not be enough.

The effort to keep New Orleans alive does not need to be only a struggle against nature; it can be a venture to cooperate with nature. The city is inextricably part of the Gulf Coast wetland ecosystem, and that ecosystem has historically operated to insulate the city against a number of dangers. The area of delta marsh that becomes open water every year is variously estimated at 60–100 km^2 (25–50 mi^2) (Shallat 2000), but Katrina destroyed 250 km^2 (100 mi^2), and chopped up the Chandeleur Islands, a barrier island chain (Hiles 2007). Many planners advocate a regional system of coastal and wetland restoration (National Research Council 2006). This would involve diversion of some of the flow of the Mississippi so that sediment can build up wetland where it has been lost (Streever 2001). It would mean establishing new protected areas and limiting development within them, especially the excavation of canals. Cypress trees would be protected, including a ban on cutting them for chipping into mulch for suburban gardens, which is today the most common form of exploitation. The barrier islands could be strengthened, extended and vegetated.

Such a program would be very expensive, although perhaps not compared with the damage that has occurred by the neglect of the ecosystem.

When I first completed this paper, it ended at this point. But on April 20, 2010 an event occurred that demonstrated that New Orleans and the Gulf Coast ecosystem are liable to disasters due to human error. The Deepwater Horizon, an offshore rig operated by Transocean Ltd. for BP (British Petroleum), while drilling for petroleum at a depth of about 5,000 ft (1,500 m) exploded and caught fire, killing 11 and injuring 17 of 126 men aboard and starting a massive oil leak from the sea floor 1,500 m below the surface. As of April 30, it has been estimated, about 62,000 barrels a day (9,900 m^3/day) were pouring upward to the surface, although at the time the volume of the flow was greatly underestimated. Efforts to plug the well, burn off the oil, or stop the spread of the oil slick, did not initially succeed. Oil had began to reach the shore from the site, which is about 64 km southeast of the mouth of the Mississippi River, and impacted many coastal wetlands, including areas that are part of coastal restoration projects, such as the Chandeleur Islands and the Delta and Breton National Wildlife Refuges. In danger were fish spawning grounds, shrimp, and oysters in the richest seafood-producing area in the US, and a large area of the Gulf of Mexico was closed to fishing and shrimp harvesting, and losses to the industry were catastrophic.

By June, in spite of containment booms that were often overtopped by high waves, oil and/or tar balls had washed up on beaches in Louisiana, Mississippi, Alabama, and the Florida panhandle. In July, oil was present in Lake Pontchartrain. Countless numbers of birds could be killed; Louisiana is visited by 70% of US waterfowl and all 110 species of migratory tropical songbirds (Tangley 2010). This was the peak time of the year for fish spawning and bird nesting. Also threatened are sea turtles and mammals including sperm whales and dolphins. Indeed, thousands of dead birds and sea mammals were collected, and many still alive were treated for removal of the oil with varying degrees of success. In addition, vegetation may be smothered and marshland turned into open water. The well was successfully capped on July 15; just before that the flow of petroleum had declined marginally to 53,000 barrels/day (8,400 m^3/day). The total volume of the spill up to the time of containment was about 4,900,000 barrels (780,000 m^3).

In order to reduce the amount of oil on the surface, BP had for a time treated it with millions of gallons of chemical dispersants, which also are toxic. It was discovered that much of the oil, along with dispersants, constitutes a "plume" under the surface of the sea, with dangers to marine life (Gillis 2010).

In its totality, this third disaster questions the safety of offshore drilling. It also exacerbates the destruction of the natural shelter offered by wetlands to the city of New Orleans and much of southern Louisiana that has been discussed in the preceding account (Burdeau and Mohr 2010) (Fig. 2.2). The total effects of the oil spill can only be estimated, and must be carefully studied, but probably will be measurable for several decades (Biello 2010).

Fig. 2.2 Images of a natural disaster—Katrina

References

Biello D (2010) Lasting menace: gulf oil-spill disaster likely to exert environmental harm for decades. Sci Am 303(1):16–18

Burdeau C, Mohr H (2010, 30 April) Associated Press

Colten CE (2003) Cypress in New Orleans: revisiting the observations of Le Page du Pratz. La Hist 64(Fall):463–477

Colten CE (2005) An unnatural metropolis: wresting New Orleans from nature. Louisiana State University Press, Baton Rouge, pp 26–28, 31, 73–75, 142

Colten CE (2006) Paradise lost? In: Davis DE, Colten CE, Nelson MK, Allen BL, Saikku M (eds) Southern united states: an environmental history. ABC-CLIO, Santa Barbara, pp 183–221, 206

Conner WH, Doyle TW, Krauss KW (eds) (2007) Ecology of tidal freshwater forested wetlands of the southeastern United States. Springer, Dordrecht, pp 448–450

Curtis SA (ed) (2007) Hurricane Katrina damage assessment: Louisiana, Alabama, and Mississippi ports and coasts, vol 3. American Society of Civil Engineers, Reston, pp 98–101

Daly EM (2006) New Orleans, invisible city. Nat Cult 1(2)(Autumn):133–148, 135

Dennis JV (1988) The great cypress swamps. Louisiana State University Press, Baton Rouge, p 2

Federal Emergency Management Association (FEMA) (2004, July 23) Release No. R6-04-093

Gillis J (2010, May 18) Giant plumes of oil forming under the Gulf. New York Times

Heerden IV, Bryan M (2007) The storm: what went wrong and why during Hurricane Katrina—the inside story from one Louisiana scientist. Penguin Books, New York, p 90

Hiles SS (2007) The environment. In: Bergal J, Hiles SS, Koughan F, McQuaid J, Morris J, Reckdahl K, Wilkie C (eds) City adrift: New Orleans before and after Katrina. Louisiana State University Press, Baton Rouge, p 16

Kelman A (August 2005) City of nature: New Orleans' blessing; New Orleans' curse. Slate Podcast. http:www.slate,com/id/2125346/

Kelman A (2006) A river and its city: the nature of landscape in New Orleans. University of California Press, Berkeley

National Research Council (2006) Drawing Louisiana's new map: addressing land loss in coastal Louisiana. National Academies Press, Washington, DC

Saikku M (1996) Down by the riverside: the disappearing bottomland hardwood forest of southeastern North America. Environ Hist 2(1):77–95, 88

Saikku M (2005) This delta, this land: an environmental history of the Yazoo-Mississippi floodplain. University of Georgia Press, Athens, pp 163–164

Shallat T (2000) In the wake of Hurricane Betsy. In: Colten CE (ed) Transforming New Orleans and its environs: centuries of change. University of Pittsburgh Press, Pittsburgh, pp 121–137

Streever B (2001) Saving Louisiana? The battle for the coastal wetlands. University Press of Mississippi, Jackson

Tangley L (2010, April 30) Bird habitats threatened by oil spill. National Wildlife

Chapter 3
Towards a Global History of Environment, Water and Climate

Water, Global Climate Change and Environmental Refugees

Ranjan Chakrabarti

Abstract Water history enhances our understanding about the nexus between the human and physical worlds within specific temporal and spatial settings. Groundwater levels are falling throughout northern China. In India, they are dropping 1 to 3 m (3–10 ft) a year. In India, they are falling in most states, including the Punjab, the country's bread basket. In the US, water levels are falling throughout the South Southwest. With 1,000 t of water required to produce 1 t of grain, food scarcity is closely tied to water scarcity. The world water deficit is historically recent. Only within the last half century, with the advent of powerful diesel and electricity driven pumps, has the world had the pumping capacity to deplete aquifers. Over pumping creates a false sense of food scarcity, it enables us to satisfy growing food needs today, but it almost guarantees a decline in food production tomorrow when the aquifer is depleted. The question of sustainability is very important.

Water history enhances our understanding about the nexus between the human and physical worlds within specific temporal and spatial settings. Groundwater levels are falling throughout northern China. In India, they are dropping 1 to 3 m (3–10 ft) a year. In India, they are falling in most states, including the Punjab, the country's bread basket. In the US, water levels are falling throughout the South Southwest. With 1,000 t of water required to produce 1 t of grain, food scarcity is closely tied to water scarcity. The world water deficit is historically recent. Only within the last half century, with the advent of powerful diesel and electricity driven pumps, has the world had the pumping capacity to deplete aquifers. Over pumping creates a false sense of food scarcity, it enables us to satisfy growing food needs today, but it almost guarantees a decline in food production tomorrow when the aquifer is depleted. The question of sustainability is very important.

As problems of water availability and competition are becoming acute, examples of past water management and social relations to water use have become increasingly relevant to our understanding of future scenarios of water use. Second, recently

R. Chakrabarti (✉)
Professor of Environmental History, Jadavpur University, Calcutta, India
e-mail: ranjan.chakrabarti@gmail.com

A. Mendonca et al. (eds.), *Natural Resources, Sustainability and Humanity,*
DOI 10.1007/978-94-007-1321-5_3, © Springer Science+Business Media Dordrecht 2012

many research findings demonstrating high quality scholarship on water history have been published. The value human societies place on water—for life, domestic use, economic production and spiritually—has led all civilizations to manipulate water flows. One of the earliest, if not the earliest, textually documented war was fought between the Mesopotamian city states of Lagash and Umma over a canal and its associated irrigated fields. Whether it is for food production, drinking water sanitation, transportation, electricity generation, or other material needs, access to and use of water has played defining roles in the development of human institutions.

Yet water, be it manipulated by human agency or not, is not necessarily beneficial. Large flooding can be very disruptive, as the Asian tsunami and the flooding of New Orleans have proven once more. The human interference with (ground) water can lead to very unexpected and unwanted consequences.

Humans attribute a variety of meanings to water, and this differs according to culture and time. Connections of water such as purity and transition are widespread. Classic examples are the preference of Hindu people to die near the river Ganges and have their ashes put in the river, the ritual of baptism which symbolizes the conversion to Christianity, and the body washing of Muslims to clean before praying.

The combination of material, cultural and religious uses and meanings of water has shaped water systems and has emphasized the intrinsic proximity and hybridity of the natural and human worlds. In addition to its physical properties, the movement of water through the larger hydrological cycle (and through that the global energy cycle as well) adds a unique chronological element. Water in its various forms is in constant motion through time and space—it evaporates from the oceans, falls to the ground, is absorbed into the soil or stored in frozen icecaps, and eventually returns to the oceans. The timescale for each droplet of water can differ greatly. Ground-water can be thousands of years old. The water we experience today most likely has come into contact with humans before.

The recent state of research on water confines that water history has developed into a vibrant historical subfield—one that incorporates and contributes to environmental history, urban history and the history of technology and landscape.

At the beginning of the twentieth century a major chunk of scientists believed that climate of the world had been essentially constant over at least 5,000 years. In the next 100 years this assumption fell through. The possible effects of past climatic shifts on human activities are yet to be explored. Historians have paid little attention to this aspect until in the 1950s and 1960s. A number of historian notably Braudel and Le Roy Ladurie showed some willingness to pay serious attention to the possible effects of climatic change in historical situations[1] (Braudel 1966). In a subsequent article on the 'History of Rain and Fine Weather', the *Annales* historian Emmanuel Le Roy Ladurie stated that, "the aim of climatic history is not to explain human history nor to offer simplistic accounts of this or that remarkable episode not even when such episodes prompt us with good reason to reflect upon the great disasters of history... ...(these are merely) spin off of the history of climate." Ladurie was concerned only with producing a picture of the changing meteorological patterns of the past ages (Ladurie 1988).

---

[1] See Nag (2007).

It is extremely urgent for the historians to understand that climate history is central to the recently unfolding sub-discipline of environmental history. It is also intimately connected with history of waters in more ways than one. Earth's hydrological cycle—the sun-powered movement of water between the sea, air, and land—is an irreplaceable asset that human actions are now disrupting in dangerous ways. Although vast amounts of water reside in oceans, glaciers, lakes, and deep aquifers, only a very small share of Earth's water—less than one-hundredth of 1%—is fresh, renewed by the hydrological cycle, and delivered to land. That precious supply of precipitation—some 110,000 km^3/year—is what sustains most terrestrial life (Postel 2006).

Like any valuable asset, the global water cycle delivers a steady stream of benefits to society. Rivers, lakes, and other freshwater ecosystems work in concert with forests, grasslands, and other landscapes to provide goods and services of great importance to human society. The nature and value of these services can remain grossly underappreciated, however, until they are all destroyed or gone.

Today human beings are tempted to think that their globalized and technologically sophisticated world is immune to harm from deteriorating natural systems. But there is no side-stepping human dependence on the water cycle. More than 99% of the world's irrigation, industrial and household water supplies comes directly from rivers, lakes and aquifers. Wetlands and river floodplains protect people from floods, provide spawning habitat for fish, recharge groundwater supplies, renew soil fertility and purify water of contaminants (Postel 2006).

The relationship between water and society is important. It is a complex historical and sociological problem. It is the vital plank on which social theories of civilization and state, community and collective action rest. Social and political organizations in all ages have been directly influenced by the ecology of water flows. Water resources of all kinds are never simply there, but are produced, used and given meaning by shifting social and political relationships. As Marx understood, the natural world is always 'humanized nature', providing a visible control of labour and history that shaped it. This is the stand of David Mosse (The Rule of Waters: Statecraft, Ecology, and Collective Action in South India, OUP, 2003) in one of the most important contributions to history of waters in south Asia. The book argues for the fundamental importance of systems of water control to Tamil social and cultural life. Until Mosse the focus was primarily on irrigation and its economic importance. Water flows have not only shaped social and political institutions, they have also legitimised them (Mosse 2003).

Mosse shows how medieval kings and chiefs exercised their power on gifted water flows, creating landscapes which inscribed their domination into the hydrology and thus naturalised it. He cites the example of a Maravar King throwing flowers into the flowing water of the channel that he had excavated, and their passage resolved an inter village conflict over water rights by disclosing the water's natural flow, political rule was naturalised into the drainage. It is high time to open up the frontier of a hybrid history of water and society. This will be a turning away from the dominant model of the economic history of irrigation towards a new kind of ethno-history or socio-cultural history of waters (Mosse 2003, p. 56).

Rivers and water resources have been central to the prosperity and survival of the human civilization. Environmental science has established that water has a very complex relationship with soils, plants and human needs. Indian cultural heritage also endorses this. Resource insensitive utilisation of water may lead to rapid disruption of the essential ecological processes that recharge and renew water resources and make them available perennially for the generation of plant, animal and human life. Modern water management in capitalistic economies looks upon water as a stock to be tapped. Such an approach to water resource management that views water as stock and not as a flow in the water cycle generates a misconception that through large man-made structures water resources can be augmented. However water resource cannot be created. It can be stored, diverted, used, polluted but its overall availability cannot be enhanced (Shiva 1991)[2]. The ecological understanding of water thus involves;

1. An understanding of the relationship between water and other elements of the ecosystem;
2. An understanding of the limits on water use enforced by the water cycle.

Climate history calls for an in-depth understanding of the inner connections between water resource on the one hand and deforestation, rainfall, soil erosion, climatic change, global warming, draught, famine, and various natural calamities on the other.

According to data gathered by Munich Re, one of the largest reinsurance companies, the loss of life and property due to natural disasters has been climbing since 1986. Economic losses from natural disasters during the last decade have totaled $ 566.8 billion, exceeding the combined losses from 1950 through 1989. More than 4 times as many "great" natural catastrophes occurred during the 1990s as during the 1950s (Postel 2006). Distinguishing natural calamities from a human—induced one is getting more difficult. Storm, floods, earthquakes, and tidal waves are natural events, but the degree to which they produce disastrous outcomes is now often strongly shaped by human activities. By necessity or choice, more people are living along coastlines, in floodplains, and on fragile hillsides—zones that place them in the midst of disasters. At the same time, the clearing of trees, disappearing water

---

[2] Vandana Shiva has drawn our attention to the negative effects of the engineering bias in water management. It fails to perceive the natural river flows as critical to drainage, to recharge of groundwater, to the maintenance of the balance between fresh water and sea water. The engineering bias in water use results in large projects, which produce serious ecological instabilities and generate conflicts. The Bengal delta may be a case to track this case. The Gangetic delta covers a huge area in Bengal. The delta has been formed by the action of three great rivers, the Ganges, Brahmaputra and Meghna. The Ganges in particular has remarkable social, cultural and economic significance. Writing in the last quarter of the eighteenth century (between 1786 and 1788), Ghulam Hussain Salim the author of *Riyazu-S-Salatin,* New Delhi, 1975 (translated by Abdus Samad) commented: 'Hindus have described volumes on the sanctity of this river. Considering the water sacred they fancy that bathing there washes the sins of a lifetime, especially bathing at certain ghats, like Benaras, Allahabad, Hardwar. The rich among the Hindus bring the water of the Ganges from long distance, take particular care of it, and on auspicious days, worship it. The truth of the matter is, that the water of the Ganges, in sweetness, lightness and tasteness has no equal, and the water of this river, however long kept, does not stink.' The recent scholarship on water, gender, environment and power is fast expanding. See Lahiri-Dutt (2006).

bodies, engineering of rivers, and destruction of mangroves has wiped out the natural safety cushions that healthy ecosystems provide. Consequently, when a natural disaster strikes, the risks of losses are higher (Postel 2006). The risk to life and property from this confluence of disaster-producing circumstances places a premium on preserving what remains of nature's protective infrastructure and restoring more of it where possible (Postel 2006).

The ongoing global warming is an established fact. The environmental scientists have drawn our attention to the long-term effects of this rising global temperature. The evil effects of higher temperatures are visible on many fronts. Crop-withering heat waves have lowered grain harvests in key food-producing regions in recent years. In 2002 record high temperatures and associated drought reduced grain harvests in India, the United States and Canada dropping the world harvest 90 million t, or 5% below consumption. The record-setting 2003 European heat wave contributed to a world harvest shortfall of 90 million t. Intense heat and drought in the U.S. Corn Belt in 2005 contributed to a world deficit of 34 million t (Brown 2006a, pp. 119–128, 2006b, pp. 59–78).

Since 1970, the earth's average temperature has risen by 0.8°C, or nearly 1.4°F. During this span, the rise in temperature each decade was greater than in the preceding one. Meteorologists note that the 22 warmest years on record have come since 1980. And the 6 warmest years since recordkeeping began in 1880 have come in the last 8 years. Three of these six—2002, 2003, and 2005—were years in which major food-producing regions saw their crops going down in the face of record temperatures (Brown 2006a)[3].

In a paper presented at the American Meteorological Society's annual meeting in San Diego, California, in January 2005, a group of scientists from the National Center for Atmospheric Research reported a dramatic increase in the earth's land surface affected by drought over the last few decades. They reported that the land experiencing very dry conditions expanded from less than 15% of the earth's total land area in the 1970s to roughly 30% by 2002. They attributed part of the change to a rise in temperature and part to reduced precipitation, with high temperatures becoming progressively more important during the latter part of the period. Lead author Aiguo Dai reported that most of the drying was concentrated in Europe and Asia, Canada, western and southern Africa, and eastern Australia (Brown 2006b, pp. 59–78)[4].

Snow/ice masses in mountains are nature's freshwater reservoirs—nature's way of feeding rivers during the waterless season. Now they are being threatened by the rise in temperature. Even a I degree increase in temperature in mountainous regions can strikingly reduce the share of precipitation falling as snow and can enhance that coming down as rain. This in turn increases flooding during the rainy season and reduces the snowmelt to provide rivers during the dry season (Brown 2006b, pp. 59–78)[5].

Reduced snow to feed the Yellow River flow will contract China's wheat harvest, the largest in the world. And India's wheat harvest, second only to China's, will be

---

[3] See the website http://www.earth-policy.org.

[4] Ibid.

[5] Ibid.

affected by the flow of both the Indus and the Ganges. Instances like this can be multiplied. The shrinking of glaciers in the Himalayas could affect the water supply for hundreds of millions of people. In countries like India and China, the water hoarded during the rainy season as snow and ice for discharge in the desiccated season would be reduced. There are many more mountain ranges where snow or ice regimes are changing, including the Alps and the Andes. If we continue raise the earth's temperature we risk losing these reservoirs in the sky on which great cities and farmers depend (Brown 2006b, pp. 59–78)[6].

Global high temperature and melting glaciers may lead to rising seas and wash away human settlements across the world. But rising seas are not the only threat that comes with elevated global temperatures. Higher surface water temperatures in the tropical oceans mean more water radiating into the atmosphere to drive tropical storm systems, leading to more frequent and more destructive storms. At the beginning of this essay I have mentioned how cyclones and storms are making considerable headway all over the world. The recent trends suggest that the regions most vulnerable to more powerful storms in Asia are East and Southeast Asia, including the Philippines, Taiwan, Japan, China and Vietnam, that are likely to bear the brunt of the powerful storms crossing the Pacific. Further west in the Bay of Bengal, the east coast of India and Bangladesh are particularly vulnerable. The east coast of India are particularly exposed to severe tropical cyclonic storms that bring high tidal waves and floods in the Indian and Bangladesh Sundarbans. The monsoon in this part of the world consists of a series of cyclonic depressions, which follow each other in more or less close succession up the Bay of Bengal. The late October cyclones are examples of the most intense tropical storms. Such cyclonic storms frequently resulted in the flooding of this region in 1909 and again in 1919. The cyclone of 1919 features prominently in the forest department records. In 1919, the authorities attributed the increase in the number of men killed by tigers to the cyclone which made natural food scare (Chakrabarti 2001). Instances of such cyclones are numerous in the official records of colonial and post-colonial India. South Asian cyclones now await their historians.

The modern world has extensive experience with political and economic refugees. We are now experiencing a swelling flow of refugees driven from their homes by environmental pressures. It has been suggested by experts that the largest potential displacement may come in low-lying Bangladesh, where even a 1 m rise in sea level would not only inundate half of the country's Riceland but would also force the relocation of 40 million people. In a densely populated country of 142 million people, internal relocation would not be easy and the worst affected will be neighbouring India (Brown 2006b, pp. 114–117). The same outcome may take place in cases of more intense cyclones in Bangladesh and the east coast of India. The fundamental factor behind the rising seas is indeed the rising global temperature. The refugee flows from declining water tables and expanding deserts are also beginning (Brown 2006b, pp. 114–117). How large these flows and those from rising seas will become is not easy to predict.

---

[6] See the website http://www.earth-policy.org.

Climatic changes through a long *duree* period and their impact on the rise or decline of civilizations are now worth looking into. Rising or falling temperatures, monsoon behaviour, melting of snow on the mountains, rising sea levels, more powerful storms and cyclones may have a message to convey regarding the interactions of the humans with the natural world. In South Asia in particular climate had been central to the growth or prosperity of human civilizations. It was most crucial to rice production or settled agriculture. We have the images of Gods like Indra or Varuna who are supposed to be in control of rain and water or climate. We are also aware how the change of climate and decline of monsoon in Northwestern India possibly led to the fall of the Indus Valley Civilization. Recent research suggests that drought, rising temperatures and desertification led to its collapse. A group of scientists at IIT Kharagpur and fellow American scientists have analysed monsoon behaviour over thousands of years through geological studies and connected it to archaeological findings (The Telegraph 2006).

They say that changes in the Indian monsoon over the past 10,000 years may explain the spread of agriculture in South Asia as well as the rise and fall of the Harappan civilization (The Telegraph 2006). Again in different periods of Indian history we have references to rulers and Kings dishing out to the peasants, hit by severe droughts, the package of exemption of land revenue.

Ancient India made an important contribution to science. In ancient India religion and science were linked together. Astronomy and astrology made progress, because the planets came to regarded as Gods, and their movements began to be closely observed (Sharma 1990). I will argue that study of astronomy and astrology in ancient India became imperative because of their connection with changes in seasons and weather condition which were important for agricultural activities. One of the most notable scholars of astronomy was Aryabhatta (fifth century) and Varahmihira (sixth century). Aryabhatta discovered the causes of the lunar and solar eclipse. He concluded that the sun is stationary and the earth rotates (Sharma 1990). It would be plausible to argue that even the progress of Mathematics in India was prompted by the need to understand the climate behaviours.

Richard Grove has already shown the connections between drought in South Asia and the institutional responses to it and the beginnings of modern scientific understandings of climatic teleconnections between global scale tropical circulation and the strength of the Asian monsoon. In colonial India, repeated droughts and famine sparked off institutional innovations and a Famine code. It also triggered off a sharp debate about the connection between deforestation and rainfall change among the British officials. It is necessary to historicize the debate in the light of the pre-colonial traditional perceptions of climate as it existed in India and South Asia. This may not be very well documented. But one has to understand that there are scattered and indirect references to climate and rainfall, pattern of tropical monsoon and cyclonic depressions in many ancient and medieval texts like *Mahabharata* and *Ramayana*, Chinese accounts, writings of Mughal court historians. The ancient Greek sources relating to Alexander's invasion may be of some help too. Cultural representations of natural calamities or climate as found in literature or various visual arts are also worth looking into. It is high time to embark on a project of constructing a comprehensive history of climate in South Asia.

References

Braudel F (1966) The mediterranean and the mediterranean world in the age of phillip II, vol 2. Harper and Row, New York (1973, Emmaneuel Le Roy Ladurie, times of feast, times of famine: a history of climate since the year 1000. Paris)

Brown LR (2006a) Outgrowing the earth: the food security challenge in an age of falling water tables and rising temperatures. W. W. Norton & Company, New York, pp 119–128

Brown LR (2006b) Plan B 2.0: rescuing a planet under stress and a civilization in trouble. W. W. Norton & Company, New York, pp 59–78, 119–128

Chakrabarti R (2001) Tiger and the Ra: ordering the Maneater of the Sundarbans 1880–1947. In: Chakrabarti R (ed) Space and power in history: images, ideologies, myths and moralities. Penman, Kolkata, pp 66–80

Earth-policy. http://www.earth-policy.org

Ladurie ELR (1988) History of rain and fine weather. In: Aymard M, Mukhia H (eds) French studies in history I. The Inheritance, Hyderabad, pp 192–214

Lahiri-Dutt K (ed) (2006) Fluid bond: views on gender and water. Stree Publications, Kolkata

Mosse D (2003) The rule of waters: statecraft, ecology, and collective action in South India. Oxford University Press, New Delhi

Nag S (2007) Rain, rain do not go away: history of rainfall, deforestation and water scarcity in Cherrapunji, the wettest spot in the globe. In: Chakrabarti R (ed) History of waters in South Asia. Sage, New Delhi

Postel S (2006) Safeguarding freshwater ecosystems. In State of the world 2006: a worldwatch institute report on progress toward a sustainable society. W. W. Norton & Company, New York, pp 41–60

Sharma RS (1990) Ancient India. Oxford University Press, New Delhi, p 203

Shiva V (1991) Ecology and the politics of survival. Zed Press, London, pp 183–188

The Telegraph, Kolkata (30 April 2006) p 1

Chapter 4
The Contribution of the Barcode of Life Initiative to the Discovery and Monitoring of Biodiversity

Filipe Oliveira Costa and Pedro Madeira Antunes

Abstract Biodiversity has been fundamental to sustain the human population, which is currently estimated at nearly 7 billion people. However, less than one fifth of the extant species are known to science, and among those only a minuscule proportion was described in any biological detail. This huge gap in our knowledge of biodiversity is in deep contrast with the extraordinary level of scientific and technological development that modern society has reached. How can we take advantage of the technology currently available to detect the putative high rates of biodiversity loss? How can we efficiently manage our ecosystems and biological communities if we do not even have a comprehensive inventory of biodiversity to start with?

The Barcode of Life Initiative (BOLI) aims to contribute to resolve these questions by building a new system for species identification using DNA sequences from standardized regions of the genome—DNA barcodes. Once fully implemented, this novel system will greatly facilitate the access to taxonomic knowledge globally and revolutionize our ability to rapidly and rigorously identify life forms in a multitude of scenarios.

We anticipate major contributions of DNA barcodes for biodiversity research when integrated with other ongoing technological, organizational and conceptual developments. This can be illustrated by the growing capacity to monitor biodiversity, which has lead to the recognition of cryptic species, their prevalence and distribution patterns. The coupling of DNA barcoding with next generation sequencing will enable to capture the structure and dynamics of complex communities with unprecedented degree of detail. This can catalyze the rate of species discovery globally and contribute to improve the way in which we conserve biodiversity.

F. O. Costa (✉)
Departamento de Biologia, Universidade do Minho, CBMA – Centro de Biologia Molecular
e Ambiental, Campus de Gualtar, 4710-057 Braga, Portugal
e-mail: fcosta@bio.uminho.pt

P. M. Antunes
Department of Biology, Algoma University, 1520, Queen Street East,
Sault Ste. Marie, ON P6A 2G4, Canada
e-mail: antunes@algomau.ca

A. Mendonca et al. (eds.), *Natural Resources, Sustainability and Humanity,*
DOI 10.1007/978-94-007-1321-5_4, © Springer Science+Business Media Dordrecht 2012

4.1 Introduction

It has been nearly 8 years since a new approach to catalogue eukaryotic life (i.e. DNA barcodes and the Barcode of Life Initiative, BOLI) was proposed and launched. Over this period a number of review and opinion papers (and also book chapters) have been produced about this subject. Most of these reviews use a relatively elaborate technical language and are directed primarily towards specific audiences, active in multiple fields associated with the general area of biodiversity (e.g. taxonomy, systematics, evolutionary biology, ecology, agriculture, human health, etc.). Somewhat detailed and narrow in scope and audience, part of the existing literature tends to focus on particular slots of life or end-users of the approach (e.g. Besansky et al. 2003; Schander and Willassen 2005; Valentini et al. 2009b). An important number of these papers discuss the rationale and scientific merits or downsides of DNA barcoding, many reflecting a controversy that has emerged from the very beginning (e.g. Moritz and Cicero 2004; Ebach and Holdrege 2005; Schindel and Miller 2005; Dasmahapatra and Mallet 2006; DeSalle 2006; Rubinoff 2006; Padial and De La Riva 2007; Goldstein and DeSalle 2010; Teletchea 2010). Finally, as a scientific initiative, DNA barcoding has also been a subject of study for social scientists (e.g. Ellis et al. 2010).

Today's dynamics and speed of scientific activity is astonishing. In its 8-year span, BOLI has experienced many salient developments, which have been increasing in recent years. In our opinion, DNA barcoding, together with other related initiatives, is revolutionizing the "business" of cataloguing life. In doing so, there are important global impacts in the biological sciences, both from operational and scientific perspectives, which need to be considered. Consequences for science may take longer to become apparent. However, in our view, despite being still in its infancy, BOLI has already influenced biodiversity research in many ways, making a broad evaluation of its contribution timely.

Talks about DNA barcoding often raise a mixture of reactions in the audience, ranging from very skeptical to enthusiastic. However, from our experience, not in a single occasion has the topic been ignored. We frequently encounter professionals working in the biological sciences, or other scientists from various disciplines, that are not aware of the DNA barcoding approach and are very curious about it. Others already heard about it, but still lack an in-depth perspective about the goals, rationale and operation. A number of these colleagues are just happy that they are not drowned in technical jargon and are able to have a global perspective of the main features of the approach, which enables to build an informed opinion about it.

In this chapter we intend to minimize the use of technical jargon and produce a text readily accessible for those less familiar with the scientific field of taxonomy. The target audience is scientists and professionals that may not necessarily be planning to use this approach, but want to be informed and updated of a major initiative with large potential impact on the biological sciences. Because the chapter is a contribution integrated as part of the Small Meeting of the International School Congress, high-school teachers, college professors as well as their respective students are anticipated audiences.

Regarding the content, we will give particular focus to the relevance of DNA barcoding for taxonomy and, consequently, its current and prospective impacts on biodiversity research and management. The approaches used by the scientific community to study biodiversity are undergoing profound changes. Although the merit of any modifications aimed at improving biodiversity knowledge is subject to strong dispute, the changes themselves seem inevitable, since many of them are driven by technological developments. We perceive DNA barcoding as one of the important catalysts of such changes.

Since taxonomy is a structural discipline within the biological sciences (Wilson 2004), any major change it experiences is prone to having large repercussions across the entire scientific field. For a non-specialist audience the way taxonomy operates may not be obvious. As such, to adequately illustrate our perceived changes in this discipline, we must first provide a brief overview of how it works and what major impediments it has been facing. It is not our goal to speculate on long-term outcomes regarding operational or conceptual changes in the field of taxonomy, but to provide a snapshot of past and current challenges.

The success of many foreseen applications of DNA barcoding depends on rigorous taxonomic knowledge and, for that reason, on effective and parallel developments in taxonomic science. Assuming such development, a myriad of applications of DNA barcoding is anticipated under various scenarios where facilitated identification of specimens is of great need. It is not our intention to provide and exhaustive list of applications, but to focus on those that may improve our ability to monitor and understand biological diversity and ecosystem functioning.

4.2   Taxonomy: Entering a New Age

4.2.1   Status Report and Challenges

According to a recent study, the total number of extant species in the world, including bacteria and other prokaryotes, is estimated at approximately 11 million species (Chapman 2009). In contrast, the number of formally known species is only about 1.9 million. This number is also an estimate, since there is no central repository for counting all the new species being described. Furthermore, many species names are in fact synonyms due to redundancy in descriptions (Patterson et al. 2010). Two remarkable conclusions are evident from these numbers: we are essentially ignorant about the largest portion of biological diversity, and we do not even accurately know the bit that we were able to describe and are thus supposed to know.

Taxonomy is the scientific discipline responsible for organizing the foundations of knowledge about biological diversity, and upon which other disciplines in the biological sciences depend. For those unfamiliar with past and current scientific approaches for cataloguing life—including not only the lay citizen but also many scientists across various fields—our startling ignorance about biological diversity may be seen as an apparent failure of taxonomy to fulfill its mission. This is in great contrast to other internal viewpoints for which taxonomy is "one of the crowning achievements of modern science" (Godfray 2002).

To be able to appreciate the contribution of taxonomy to science, we first need to understand the challenges posed by the inherent complexity of biological diversity. There are extraordinary conceptual and operational challenges. Conceptual difficulties start with the concept of species. Although generally considered the prime matter of taxonomy, to this day there is no general agreement among researchers about what defines a species. Part of the problem stems from the absence of a universally applicable concept that could embrace all biological complexity and diversity. The discussion on this topic is far beyond the scope of this chapter, so we refer the readers to other studies (e.g. Hey 2006; de Queiroz 2007). Suffices to say that this has been a conceptual challenge fostering fragmentation of the taxonomic community and, therefore, leading to operational difficulties (see for example Padial and de la Riva 2010).

Fragmentation of the scientific taxonomic community has also been driven by the inability to find universal criteria for species recognition. Since Linnaeus that morphological and anatomical features have been the prime traits used for the establishment of species boundaries, particularly for animals and plants. Nonetheless, the morphological nuances that enable species level discrimination are often so intricate that few experts globally are able to tell them apart accurately. Few people realize how hard it can be to discriminate species, and this is perhaps one of the reasons for the underestimated value of the research undertaken by taxonomists.

Due to the massive morphological diversity among organisms, taxonomists were forced into increasing taxa-specific specialization. The isolation of taxonomists within the collegial community of their target taxa, and the atomization of the whole discipline was inevitable. The number of specializations grew significantly and the ability to exchange information decreased. The problems with fragmentation extended also to the taxonomic scientific literature (Moritz 2002; Knapp et al. 2007), creating serious operational problems. It lead to a situation where being a taxonomist meant dominating the history, sources and details of the literature about a particular taxonomic group.

Surely taxonomists were able to build a consistent code of rules to organize taxonomic knowledge, such as for example the International Code of Zoological Nomenclature or the International Code of Botanical Nomenclature. However, only recently a centralized web-register for animal species names was proposed and implemented (i.e. ZooBank; Polaszek et al. 2005). Biological specimen collections were created and nurtured, and important infrastructures built to harbor them, such as natural history museums and herbaria. It could be that lay citizens perceive Natural History Museums (NHM) solely as an organized display of natural history for the cultural benefit of societies, just like an art museum. Few will be aware of the other equally important facet of NHMs—the one of libraries of biological specimens where actual research is conducted. This perception could change dramatically with a visit to the backstage of a NHM.

The fact that, on average, 18,000 new species are described annually (Chapman 2009) demonstrates the effectiveness of the work by taxonomists, and the supporting body of infrastructures and approaches created. In spite of the merits of the hitherto organization and progress of taxonomic science, the available resources and approaches are still far from being able to counter the fundamental challenge of cataloguing all biological diversity; a task of tremendous complexity and massive scale. This problem was finally perceived and assimilated by the scientific community not long ago, and became commonly referred to as "taxonomic impediment" (Rodman and Cody 2003). Hence, the problem of the taxonomic impediment runs in parallel to what many claim as a gradual decline of the impact of taxonomy on the global scientific arena and the support of society (Godfray and Knapp 2004; Wheeler et al. 2004; Boero 2010).

Much has been discussed on the sources and solutions for the taxonomic impediment (Rodman and Cody 2003; Godfray and Knapp 2004; Lyal and Weitzman 2004; Raven 2004; Wheeler et al. 2004; Agnarsson and Kuntner 2007; de Carvalho et al. 2007; Evenhuis 2007; Knapp et al. 2007). Causes for the decline of taxonomy go as far as science policy and negative discrimination by funding agencies to what is perceived solely as descriptive science as opposed to more fashionable hypothesis-driven research (but see Wheeler 2004). The current metrics of scientific output, as for example journal impact factors, is often negatively biased against taxonomists and taxonomic work (Bouchet 2006; Agnarsson and Kuntner 2007; Boero 2010). Since these ambiguous metrics determine assess to funding and even employment opportunities, taxonomists are more affected by their growing prominence in scientific curricula than researchers in other fields. Conceptual challenges in taxonomy remain unresolved (e.g. species concepts, Wilson 2003) and new approaches may add complexity to an already turbulent debate. On the operational front there appears to be more consensus about the perils of fragmentation. It is at this level that major improvements are anticipated. Appeals for unity within the discipline and integration of approaches and tools multiply (Godfray and Knapp 2004; Wheeler et al. 2004; Dayrat 2005; Goldstein and DeSalle 2010; Padial et al. 2010; Schlick-Steiner et al. 2010). Taxonomy needs to improve cohesion and find unifying platforms to enhance its contribution and efficiency.

4.2.2   Signals of Renaissance

As the debate continues, exceptional technological developments are boosting at a global level in multiple areas (e.g. technologies associated with genome analysis and manipulation, organization and transmission of information technology (IT), imaging technology, the dissemination of Global Positioning Systems (GPS) and Geographic Information Systems (GIS)). Among these, we will focus on two that we consider exceptionally capable of having significant effects on the mode of operation of taxonomy (i.e. genomics and IT). Technology is not expected to be the only solution to the problems of taxonomy, but it can improve much of its operational inefficiencies and revive the whole discipline to a new level of impact on the global scientific arena and the way it translates into services to society.

DNA-based approaches have unique features both as a source of information to understand biodiversity and as a tool for taxonomists. To start with, it provides a set of universal characters directly comparable among very different organisms; a major advantage compared to the morphological approaches discussed earlier. This considerably facilitates comparisons, validations and verifications among experts. Moreover, it enables access to non-taxonomic experts, like many ecologists that require rigorous species identifications in their research.

Another important asset of DNA-based approaches is the ability to access biodiversity beyond morphology, down to the molecular level. Such capability is critical for example to diagnose and study cryptic species (organisms morphologically identical or nearly-identical but actually representing distinct species, which are evolving independently), identify species from body parts, and to establish definitive correspondence between larvae and adult morphotypes and sex morphotypes (e.g. see particularly striking example in Johnson et al. 2009).

With the growing utilization of DNA technology to study biodiversity and explore its benefits at a biotechnological level, DNA and tissue banks started to be built either as separate initiatives (e.g. The DNA Bank Network, http://www.dnabank-network.org; Ocean Genome Legacy, http://www.oglf.org/) or within institutions holding public biological collections. These DNA banks constitute novel and important repositories for the study biodiversity, which can potentially provide access to information contained in the genome of millions of specimens. However, the access and type of use given to biodiversity resources (including DNA banks) can become a source of contention, because biotechnological exploration and commercial gains are possible. To this end, the Convention for Biological Diversity (CBD) initiated the process of establishing an international legal framework to ensure that countries providing access to their biodiversity data will also benefit from eventual commercial exploitation (International Regime on Access and Benefit Sharing; Schindel 2010). Because this regime only applies to research aimed at developing commercial products, many researchers may face growing difficulties to access biological specimens, tissues and DNA (Bouchet 2006; Schindel 2010).

Contrary to nucleic acid-technologies, the role and potential of IT for the progress of taxonomic science appears to be consensual. Currently, IT is a vital resource for taxonomy. Nevertheless, a substantial part of the taxonomic impediment resides on the difficulties to manage, share and access outstanding amounts of relevant information, dispersed through many institutions and documentation (Moritz 2002). Among the numerous uses of IT for taxonomists, perhaps the development of dedicated structured databases and the ability to share biodiversity information and documentation on the Internet are the most influential tools. From very early on, taxonomists initiated the development of a diverse set of databases for their benefit. Databases can be dedicated to particular taxonomic assemblages, geographically oriented or service oriented, among many other types. Of particular relevance are databases of NHMs and biological collections that facilitate crucial information for taxonomists, such as location and quantity of specimens available for examination, particularly type specimens.

Numerous taxa-oriented databases have been developed, sometimes in a dispersed and *ad hoc* fashion. The content of taxa-oriented databases varies from as little as providing purely taxonomic information (i.e. lists of species names, authorities, systematic position and synonyms) to as much as including images, maps with actual and estimated geographic distribution, and entries with biological and ecological details (e.g. FishBase; Froese and Pauly 2010). With such a large profusion of different types of taxonomic databases, one might think that dispersal would become a problem. It is not necessarily so. Informatics provides straightforward solutions for dispersal by enabling database interactivity and the creation of integrated databases. Databases such as the Catalogue of Life (http://www.catalogueoflife.org/info/about, established in 2001 to merge Species 2000 and the Integrated Taxonomic Information System (ITIS) catalogues), or the World Register of Marine Species (WoRMS, Appeltans et al. 2010) provide centralized access to baseline taxonomic information about a range of extant organisms, being fed of original data by smaller scope databases (95 taxonomic databases in the case the Catalogue of Life, including WoRMS). The CBD also adopted a similar biodiversity information aggregator website: the Global Biodiversity information Facility (GBIF) an "internet-based index of a globally distributed network of interoperable databases that contain primary biodiversity data".

Regarding biodiversity information, one of the major recent developments and perhaps the most influential for the future, was the creation of the Encyclopedia of Life (EoL) in 2008 (Wilson 2003). This initiative can be described as an "ecosystem of biodiversity websites". EoL's main goal is to provide a global cluster for detailed and comprehensive information on every species on the planet. Rather than generating original data, this portal supplies structured information fed in by dynamic links to data-generator databases. Therefore, EoL takes advantage of existing biodiversity information to create a much-awaited one-stop access point for detailed information about all living organisms. An important component of EoL is the Biodiversity Heritage Library, a project aimed at providing free web access to the original literature of each species (http://biodivlib.wikispaces.com/About).

Even though efficient solutions to the large data dispersal could be found, two main problems of taxonomic databases emerged: (1) conflicting data among databases; and (2) a more latent and conspicuous problem of unstable funding for maintenance and development. Many databases were developed based on voluntary efforts by individual researchers or small communities of taxonomy professionals. Their maintenance is dependent on time availability of those involved and frequently on temporary funds. In the genomics arena the creation of multiple global repositories of genetic information (e.g. GenBank, EMBL or DDBJ) was fundamental for the development of this technology and for its successful implementation in the practice of many research areas (e.g. biomedical research, biotechnology). Public outreach and awareness are key elements in driving political will and funding allocations. Because the actual and potential benefits for the economy and society became obvious for lay-citizens and policy-makers, the genomics community was able to gather much support for the creation of rather permanent IT infrastructures to assist in research. A number of countries decided to fund public databases of genes and proteins (e.g. SWISS-Prot, a curated protein sequence database, http://expasy.org/people/swissprot.html), which were linked through large integrated networks, thereby becoming vital resources for researchers.

The example of genomics demonstrates that it is possible for discovery-based science (or at least perceived as such) to gather large-scale support if downstream applications and societal benefits are properly marketed to the public. For taxonomy to thrive, similar capability to integrate efforts and gather support is needed. Taxonomy could gain inspiration from genomic databases and some of their approaches to curate data. The concept of annotation of DNA sequences and genes for instance, could be particularly useful for taxonomy if properly transferred to the realm of biodiversity.

Indeed this concept is already being implemented in the operation of the EoL, partially inspired by the model of Wikipedia. In addition to institutional partners, individual scientists or any other citizen is invited to participate in the completion of the encyclopedia and to contribute with various types of information to EoL. Scientific reliability of the data is vetted by a network of data curators, which are experts in particular taxonomic groups and are responsible to validate or comment on data entries. If the scientific community eventually adopts this tool, such a wiki-type model of interaction among taxonomic experts will enable much faster revision of problematic species complexes and clarification of many inaccuracies and conflicting information spread across non-curated websites.

The Global Taxonomy Initiative (GTI), created by the Convention for Biological Diversity calls for the need of more human and logistic resources devoted to taxonomic research. It does so by taking a systematic approach driven by a clear political will. Particular emphasis is placed in improving research conditions for regions of the globe with poor taxonomic infrastructure. Some countries took active and explicit measures to solve the taxonomic impediment. In the USA a multi-annual program was launched to train a new generation of taxonomists (Partnerships for Enhancing Taxonomic Expertise in Taxonomy—PEET) (Rodman and Cody 2003). In Europe, a Consortium of European Taxonomy Facilities (CETAF) was established in 1996, which later gave origin to the EU-funded network of excellence the European Distributed Institute of Taxonomy (EDIT, http://www.e-taxonomy. eu/). More recently, a European program to develop virtual research communities involved in biodiversity research was launched (Virtual Biodiversity Research and Access Network for Taxonomy—ViBRANT, http://vbrant.eu/).

4.2.3 Renovation of Taxonomy

Taxonomy is undergoing a profound change and coming to a new age (Godfray and Knapp 2004; Wheeler 2008). Time has come for a number of parallel technological developments and global initiatives to meet and contribute for its renaissance. It is now possible for the taxonomic community to embrace and articulate multiple technologies (e.g. DNA technology, imaging, computing and information, GPS) for the benefit of gaining more complete and deeper knowledge of biodiversity. Large-scale initiatives such as EoL are expected to have a profound impact to achieve these goals and to overcome important operational constraints.

With a new organizational framework and extensive use of these novel tools, taxonomy may have increasing influence in society, delivering faster and improved services that are currently in great demand (see Sect. 4.4). Society may in turn gain better awareness of the value of taxonomic infrastructures and demand support for their maintenance and development. Growing awareness of global environmental problems such as climate change and biodiversity loss will drive these changes even further (Bouchet 2006).

Taxonomy, taxonomists and taxonomic infrastructures may regain a central role in the biological sciences. The next generation of taxonomists will need to master the use of a variety of technologies and integrate data from different sources and approaches into a more holistic understanding of biodiversity. Crucial for these developments are the global initiatives and large-scale resources able to aggregate the taxonomic community around flagship projects. As we discuss in the next section, the BOLI, and iBOL in particular, will have an important share of contributions to this end.

4.3 Synopsis and Update on the Barcode of Life Initiative

4.3.1 Brief History and Synopsis of Main Developments

Less than 10 years ago an article by Paul Hebert and colleagues proposing the use of DNA barcodes in species identifications was published in the *Proceedings of the Royal Society of the United Kingdom*. Hebert et al. (2003a) proposed the use of short (i.e. about 650 base pairs) DNA sequences from one pre-defined region of the eukaryotic genome (e.g. the mitochondrial gene cytochrome *c* oxidase subunit 1, COI) as a molecular tag for species identification in any taxa. The authors claimed two major reasons and goals for the application of the DNA barcode concept: (a) simplification and facilitation of species identification in a multitude of scenarios, so that it could more easily be accessible to non-experts, and, simultaneously, (b) speeding up the discovery of the massive number of unknown species, thereby contributing to overcome the taxonomic impediment.

In 2003 the application of molecular genetic markers in species diagnosis was not novel. The use of molecular markers had been introduced decades earlier, and experienced great expansion subsequently (Carvalho 1998). The proposal by Hebert et al. (2003a) differentiated from previous practice due to its degree of standardization and scale: a single DNA barcode sequence region for all eukaryotic species. Technological developments in DNA technology enabled easy access and the routine use of DNA sequencing in many research laboratories, serving as a catalyst for such an ambitious proposal.

But there were more attributes of DNA barcoding that differentiated it from other molecular-based proposals that emerged concurrently, as for example DNA taxonomy (Tautz et al. 2003; discussed in Teletchea 2010). DNA barcoding is intended to integrate within the existing taxonomic practice as an additional tool available for taxonomists (see Goldstein and DeSalle 2010), although with special strengths and potential. Hence, rather then trying to cast aside the current taxonomic practice and infrastructure, DNA barcoding may contribute to improve its output (Hebert and Gregory 2005; Schindel and Miller 2005). Nature's cover page of 15 March 2007 featuring a drawing of Linnaeus in the field wearing modern outdoors clothing while holding a barcode could not be more illustrative.

In Table 4.1 we summarize a selection of what we consider the most relevant developments of the BOLI since it was first proposed. We include key debates that appeared in the literature, because they were crucial to clarify the concept and operation of DNA barcoding, and also for its influence in the taxonomic practice.

DNA barcoding had one great merit: the one of stimulating debate among the scientific community. This goes hand in hand with the actual contribution to advance the knowledge of life. Forcing the debate is perhaps a more accurate way of describing it. After all, DNA barcoding was proposing a new global standardized system for species identification that claimed to help countering the taxonomic impediment: no professional in the discipline could ignore it.

By boosting the debate around this new approach, DNA barcoding drove scientists, particularly taxonomists and ecologists, to re-think and actively discuss the taxonomic impediment and the whole enterprise of cataloguing life and how it could get closer to societies' needs. The debate has been rich, with many angles of the problem brought into discussion, from those purely scientific, to more operational concerns, all the way to science policy (e.g. debates published in *Systematic Biology*; Hebert and Gregory 2005; Smith 2005; Will et al. 2005 and in *Genomics, Society and Policy*, Costa and Carvalho 2007a, b; Dupré 2007; Hollingsworth 2007; Holm 2007). To this day the controversy around DNA barcoding did not end, and some of the arguments are so extreme that there is little leeway for any meaningful conclusions or benefits to emerge from them (e.g. Ebach and de Carvalho 2010).

Upstream from other discussion topics, the argument about feasibility and reliability of the DNA barcoding approach encouraged both sides of the controversy to gather evidence in support of their claims. This new common ground for scientific challenge became an opportunity to confront data obtained using different approaches, or to explore new perspectives to investigate biodiversity by combining various sources of evidence. In this process, important patterns have emerged for various groups of organisms (e.g. parasitoids, fish). Also, comparisons between morphological and molecular data significantly improved the knowledge of particular taxa.

Initially, DNA barcoding supporters focused on generating key proof-of-concept studies, demonstrating the applicability of the approach for different animal taxa (e.g. Hebert et al. 2004b; Ward et al. 2005; Hajibabaei et al. 2006; Costa et al. 2007). One of the first such studies focused on birds (Hebert et al. 2004b) and its publication generated one of the most relevant debates published in the same issue

Table 4.1 A selection of the most relevant events, debates and developments in the Barcode of Life Initiative

Event	Comment	Key references or website
Scientific article proposing DNA barcoding	Publication of the original article proposing the DNA barcoding approach by P. D. N. Hebert and colleagues. A second paper followed shortly in the same journal describing patterns of nucleotide divergence in the proposed DNA barcode region (Hebert et al. 2003b) Note that, although the term "molecular barcode" was used in an earlier article by Floyd et al. (2002), the term "DNA barcode" refers usually to the approach proposed by Hebert et al. (2003a)	Hebert et al. (2003a, b)
Establishment of the Consortium for the Barcode of Life—CBOL (April 2004)	A consortium of natural history museums, herbaria, academic institutions, governmental agencies, among other organizations, was established with the mandate of promoting and coordinating the global implementation of DNA barcoding. Since 2004, over 200 organizations from 50 countries have joined CBOL. CBOL is funded by the Alfred P. Sloan Foundation	http://barcoding.si.edu/
Identification of birds through DNA barcodes (October 2004)	One of the first proof-of-concept articles demonstrating the applicability of DNA barcodes for species identification, in the case of the well-known avifauna of North America (over 500 citations as of January 2011). A comment to this article by Moritz and Cicero (2004) was published in the same issue, promoting extensive debate about DNA barcoding	Hebert et al. (2004b), Moritz and Cicero (2004)
Ten species in one—DNA barcoding helps to sort out a complex of 10 moth species, previously thought to be only 1 species (October 2004)	This paper was an important showcase for the potential of DNA barcoding to help unravel complexes of cryptic species. In this case, analyses of DNA barcodes of a species of neotropical skipper butterfly produced 10 distinct barcode arrays. Each barcode array corresponded to caterpillars with a distinct color banding pattern and specific plant food sources, indicating the presence of 10 different species with an identical adult morphotype. After Hebert et al. (2003a) it is the second most cited DNA barcoding paper up to the present date	Hebert et al. (2004a)
Taxonomy in the twenty-first century—a themed issue of the Philosophical Transactions of the Royal Society of the UK (April 2004)	In this publication with such an iconic theme, DNA barcoding was already discussed in several papers despite little over 1 year had passed since the publication by Hebert et al. (2003a). This illustrates the immediate impact of the DNA barcode proposal on the scientific community. Some seeds for a debate that continued further were launched here (e.g. Wheeler 2004). DNA taxonomy and environmental barcoding (Blaxter 2004) were also discussed in this issue	Blaxter (2004), Godfray and Knapp (2004), Janzen (2004), Wheeler (2004)

Table 4.1 (continued)

Event	Comment	Key references or website
First international conference on the barcode of life NHM London (February 2005)	The first international conference, significantly held at the London NHM was an important milestone and a major boost for BOLI. An important deliverable of this meeting was the collection of papers published in a special issue of the Philosophical Transactions of the Royal Society of the UK, and the creation or reinforcement of global campaigns	Savolainen et al. (2005), Godfray (2006)
DNA barcoding debate at the 5th biennial conference of the partnerships for enhancing expertise in taxonomy (PEET) (September 2004)	This DNA barcoding debate at the PEET conference, and the ensuing papers published in Systematic Biology, are one of the key landmarks in the path of BOLI. Following a paper introducing the debate (Smith 2005), the opposing views on the merits and perils of DNA barcoding had the opportunity to present their cases. The scope of the debate was broad, including topics of scientific validity to science policy-related issues	Hebert and Gregory (2005), Smith (2005), Will et al. (2005)
Launch of Fish Barcode of Life—FISH-BOL (June 2005)	FISH-BOL was the first global DNA barcoding campaign to be established. The concept started in 2004 and was formally launched in June 2005 at the inaugural FISH-BOL workshop. The FISH-BOL campaign represented a major flagship project for the barcoding community, focusing on a taxonomic assemblage with a tractable number of species, with global and significant economic relevance, and many straightforward prospective applications for DNA barcoding (e.g. Costa and Carvalho 2007b). Other global campaigns followed soon thereafter, such as the All Birds Barcoding Initiative, and All Lepidoptera Barcoding campaign Barcoding campaigns became a key component of the BOLI activity and many new campaigns were launched since then, not only taxa-, but also ecosystem- (e.g. Marine Barcode of Life), geographically-, or country-oriented	Ward et al. (2009)
Proposed standards for BAR-CODE records in INSDC (BRIs) (November 2005)	The Data Working Group of CBOL published the proposal for implementation of a "BAR-CODE" keyword or flag to be included in the three major nucleotide sequence repositories (International Nucleotide Sequence Database Collaboration—INSDC), reserved for sequences meeting the data standard for a DNA barcode The BARCODE flag was implemented shortly after, and the Data Working Group document was last reviewed in 2009	Hanner (2005), Walters and Hanner (2006)

Table 4.1 (continued)

Event	Comment	Key references or website
First European DNA barcoding conference (held in Leiden, 2007) and second international barcode of life conference (held in Taipei, 2007)	In September 2007 the second international barcode of life conference (Taipei) and the conference "DNA barcoding in Europe" nearly overlapped, denoting the recent growth of the community involved in the initiative. The latter conference gave way to the establishment of the European Consortium for the Barcode of Life (ECBOL). Multiple DNA barcoding meetings followed, namely the 3rd conference in Mexico city and the 2nd conference of ECBOL (ECBOL2) in Braga, Portugal. They were characterized by a continuously growing diversity of subjects and participants, thereby reflecting increasing adherence and impact on the scientific community	http://barcoding.si.edu/cbol_taipei/ http://www.ecbol.org/
Quarantine Barcode of Life (QBOL) (May 2009)	The Quarantine Barcode of Life (QBOL) is a research project funded by European Union's 7th framework program, with the objective of developing a DNA barcode-based system for identification of potential pathogens carried by biological products through international trade before they enter Europe. It can be viewed as model project of BOLI, since it builds on the strengths of DNA barcoding to deliver a service of straightforward application and with high economic and ecological relevance in the modern global society	http://www.qbol.org/
Proposal of plant barcodes (September 2009)	The plant working group of CBOL proposed two plastid genes (rbcL and matK) to serve as the primary plant DNA barcodes. The much awaited plant barcodes took several years and numerous trial studies on plant species to generate a fair consensus. These two plant barcodes cannot discriminate per se all plant species, but they will be able to do it in the majority of the cases. The proposal of plant barcodes opened doors for a large community of plant researchers to engage in BOLI	Hollingsworth et al. (2009)
The International Barcode of Life (iBOL) formal launch (September 2010)	In September 2010, a global DNA barcoding project was eventually launched—the international barcode of life (iBOL), with the main goal of creating a library of DNA barcodes for 500,000 eukaryotic species by 2015. It will require the acquisition of DNA barcodes either collected from the wild specimens or sampled from collections. To our best knowledge, there is currently no matching global biodiversity project in terms of scale and ambition	http://ibol.org/
	The potential impact of iBOL on short to medium-term discovery and monitoring of biodiversity appears evident	

of the *Public Library of Science—Biology* that focused on the scientific aspects of the controversy (see Moritz and Cicero 2004). This debate was important to examine the limitations of the approach (e.g. recently diverged species, introgressive hybridization, among other limitations), and to initiate the clarification of the role of DNA barcoding in taxonomy: not a substitute but a complement to conventional approaches. Further clarification was pursued with a study that benchmarks the role of DNA barcoding as a complement of taxonomy, phylogeography and phylogenetics (Hajibabaei et al. 2007).

One of the earlier scientific criticisms of DNA barcoding was that sequence clusters obtained for COI (or other selected barcodes—see Table 4.1) did not necessarily match species boundaries. In a study with the bryozoan *Celleporela hyalina* (Gomez et al. 2007), a key test was undertaken going as far as the biological species concept, to check for correspondence among reproductively isolated populations, barcode clusters, and morphological variants. Largely based on the lack of morphological variability, *C. hyalina* was considered at the time to be a single cosmopolitan species, occurring in nearly all oceans. Mating trials among specimens from distant populations revealed many reproductively isolated units that matched monophyletic clusters, obtained from analyses of COI barcodes and nuclear gene sequences variation. Hence, this study worked both as a strong proof-of-concept for DNA barcoding and as a demonstration of its potential to clarify evolutionarily meaningful biological diversity hidden behind morphological homogeneity. Another similar example was found in the moth *Astraptes fulgerator*. Although the adults are indistinguishable, the caterpillars present distinct morphological patterns, feed on specific plants and separate among 10 DNA barcode species clusters (Hebert et al. 2004a).

These articles and many subsequent studies have consistently shown agreement between monophyletic units formed by barcode clusters and established species boundaries (although there are characterized exceptions). The reasons for this coincident pattern of variation are unknown, but a solution may emerge from attempts to explain evolutionary mechanisms capable of generating low intraspecific variability in DNA barcodes (see Lane 2009). Some authors suggest that mitochondria bioenergetics and, therefore, its underpinning genetic variation, might be an important driver in speciation (Gershoni et al. 2009). The massive amount of data on intra- and interspecific variability of the mitochondrial gene COI to be collected within the scope of BOLI may contribute to resolve these and other important questions about molecular evolution and speciation (Costa and Carvalho 2010).

Assuming that all or most of the publications pertaining to DNA barcoding have cited Hebert et al. (2003a) and considering that its total number of citations to date surpasses 1,000, it can be said that DNA barcoding is a "hot topic". However, a better measure of its influence in the scientific community is reflected not as much by the profusion of studies, but by the diversity of approaches, target taxa, and even the broad geographic distribution and professional interests of the citing authors (a detailed analysis is available in Teletchea 2010). Few initiatives in the history of species biodiversity had the merit of bringing together such an heterogeneous community around a common approach. This heterogeneity became apparent upon creation of the Consortium for the Barcode of Life (CBOL), which aggregated a variety of institutions, from museums and herbaria to governmental agencies. This is clearly visible through the wide array of topics and participants present at major international barcoding conferences (Table 4.1). Taxonomic experts in widely disparate groups (e.g. from phytoplankton to mammal experts) have met with molecular systematists, ecologists working on distinct topics, molecular biologists, bioinformatics professionals, etc.

Among the recent progress in the DNA barcoding enterprise, the most influential was the formal launch of the International Bacode of Life (iBOL—www.ibol.org) in September 2010. iBOL is a global DNA barcoding project aiming—among other subsidiary goals—to generate a reference library of DNA barcodes for 500,000 eukaryotic species by 2015. Given its taxonomic breadth and geographic scope, targeting all sweep of eukaryotic life and all biomes, it is very likely the largest and most ambitious biodiversity project to date. CBOL, and later iBOL, have been prone to establish partnerships with relevant international and national initiatives and institutions (e.g. International Nucleotide Sequence Database Collaboration (INSDC), GBIF, Species 2000, ITIS, EoL), hence becoming an important aggregator of biodiversity-related research (Fig. 4.1 depicts the DNA barcoding pipeline and how it relates to other parallel initiatives and institutions). The CBOL's proposal to specifically flag compliant DNA barcode records (Hanner 2005; Walters and Hanner 2006) had the merit of stimulating the rise of data quality standards for submission of nucleotide sequences to INSDC. Indeed, some authors have been alerting for the perils of inaccurate data in sequence entries in the shared INSDC dataset (Harris 2003; Bidartondo et al. 2008). The barcode flagging approach can help separating "COI-like" barcode sequences (Buhay 2009) from actual DNA barcode entries, whose validity can be scrutinized. Taxonomic identifications associated with sequence entries are mistaken more often than acceptable, and data collection pertaining to the specimens is often absent or rather incomplete. In addition, despite its relevance for future validations, the archival of voucher specimens used in molecular studies is rare (Pleijel et al. 2008). Archival of voucher specimens in publicly available collections is one of the requirements for building reference libraries of DNA barcodes.

The evidence collated above illustrates unequivocally the actual and potential contributions of BOLI to advance knowledge of biodiversity, and as one important catalyst of change and aggregation in the operational framework of taxonomy. In the following section we discuss the anticipated direct benefits of DNA barcoding for biodiversity discovery and monitoring, as well as the indirect contributions to our understanding of ecosystem functioning.

4.4 DNA Barcoding in the Discovery and Monitoring of Biodiversity

The effective implementation and utilization of DNA barcoding as a global system for species identification requires at the front end the creation of a comprehensive reference library of DNA barcodes matching each species to a diagnostic barcode

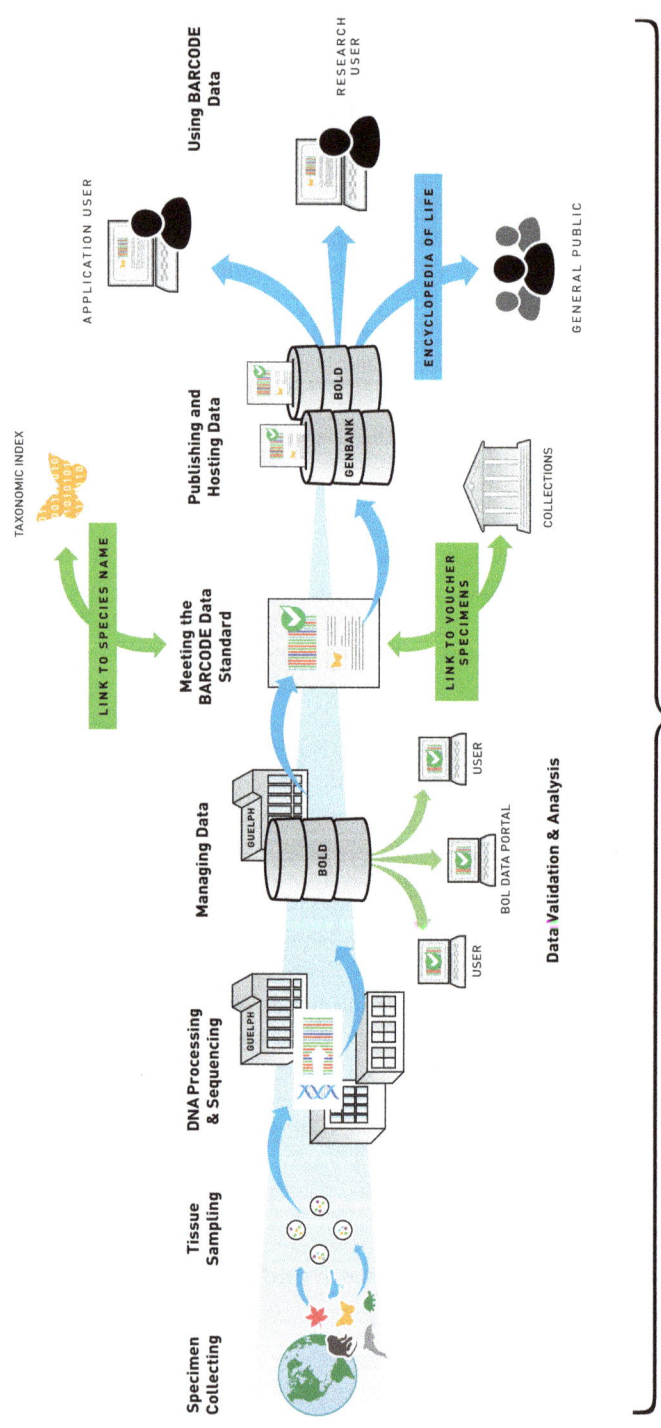

Fig. 4.1 Schematic drawing of the steps involved in the creation of reference libraries of DNA barcodes. Note the links with involved institutions such as natural history museums, and related initiatives like the Encyclopedia of Life. (With permission from the Consortium for the Barcode of Life)

sequence or array of sequences. It is assumed that this is possible for the majority of species; a small percentage may be untraceable by these methods.

The creation of the reference library is a long-term and ambitious endeavor. A significant milestone will be achieved at the end of the first phase of iBOL with a library available for half a million species. However, the full potential of the global identification system can only be completely realized with the library nearly completed, which, as we will discuss, may not be entirely feasible.

DNA barcoding can be especially valuable for the discovery of biodiversity in some realms of life. The inventory of the oceanic biome is a particularly challenging undertaking. The number of known marine species ranges from 250,000 to 274,000 (Bouchet 2006), a relatively small number compared to terrestrial life and, therefore, apparently more easily traceable. However, the marine metazoan diversity spans over 30 different phyla, whereas in terrestrial animals about 90% of the species belong to a single phylum (Jaume and Duarte 2006). With such high morphological diversity, tackling marine biodiversity demands much higher specialization and distribution of taxonomic experts among taxa. Notably, while the discovery of new higher-rank taxa in terrestrial systems is relatively rare nowadays, there are regular reports for marine organisms, including the recent discovery of new phyla (Jaume and Duarte 2006). This is a strong sign that many groups of marine organisms are still very poorly known. They generally depend on a handful of experts to describe and investigate them.

Furthermore, vast areas of the oceans such as the deep sea floor remain virtually unexplored and, despite the surprising high levels of biodiversity that have been found there, operational difficulties in reaching and studying deep-sea life will continue to hinder discovery efforts (Jaume and Duarte 2006). Under this scenario, any attempt to estimate the number of species in marine environments is often no more than an educated guess. A conservative estimate points to between 1.4 and 1.6 million species as the total marine biodiversity in the world, meaning that at current rates of description it would take 250–1,000 years to complete the inventory of marine life (Bouchet 2006). Even though, for the reasons presented above, a comprehensive inventory of marine life will remain an extraordinary challenge, the knowledge gap is still so large that even a modest contribution from DNA barcoding could lead to major advances in our knowledge of the ocean's life (see Schander and Willassen 2005; Bucklin et al. 2010, 2011).

A comprehensive reference library of voucher specimens is a pre-requisite for the implementation of DNA barcoding in the discovery and monitoring of biodiversity. Nevertheless researchers already started to test applications in particular scopes using a wide variety of methods. Some are inspired by the DNA barcoding concept, but deviated somewhat from the original protocol to accommodate features of particular target taxa. For instance, from the onset it was apparent that the application of DNA barcoding to microbiota might require a distinct approach as opposed to that followed for macrobiota. The barcoding approach for macrobiota is based on the existing framework and practice of taxonomy, including the relevance given to voucher specimens and the integration of molecular and morphological characters (Costa and Carvalho 2007a, b). This, however, may not be appropriate for the microbial world where the concept of species becomes in many cases uncertain, and genetic/morphological/metabolical diversity is so high that hinders capacity to create reference libraries of voucher specimens. For example, arbuscular mycorrhizal fungi (AMF), widespread belowground symbionts with most vascular plants, constitute one of the seven phyla in the Kingdom fungi (i.e. Glomeromycota) (Smith and Read 2008). Although approximately 230 species of AMF have been formally described and catalogued (see http://www.lrz.de/~schuessler/amphylo/), field surveys using DNA-PCR based methods consistently find many undescribed molecular operational taxonomic units (MOTUs) as well as non-saturated accumulation (or rarefaction) curves (Fitter 2005; Öpik et al. 2009). Accumulation curves are a method used to determine whether the biological diversity of a given area has been sampled efficiently (Gottelli and Colwell 2001). For a given area and taxa, multiple samples are collected and the species diversity determined. The accumulated number of taxa observed is then plotted against the number of screened samples. As the number of samples increase, these accumulation curves eventually reach and asymptote, indicating the point at which the diversity has been properly sampled. For logistical reasons associated with the biology of AMF, their taxonomy has traditionally been based on spore morphology (e.g. Sieverding and Oehl 2006). This is rather limited not just because extensive expert training is required, but because some AMF may not form spores and different species can form morphologically identical spores (Gamper et al. 2009). Furthermore, the genetic makeup of these microorganisms is complex (Pawlowska and Taylor 2004), and there seems to be extensive within species variation (Koch et al. 2006; Antunes et al. 2011), thereby challenging the concept of species. Even though international culture collections of AMF exist (INVAM, http://invam. caf.wvu.edu/; BEG, http://www.kent.ac.uk/bio/beg/; GINCO, http://emma.agro.ucl. ac.be/ginco-bel/), these soil-born symbionts can be difficult to maintain in culture, and some species described before DNA sequencing technology was developed are no longer available. A comprehensive open-access database containing Glomeromycota DNA sequence data collated from publicly available taxonomic and ecological publications has recently been created (MaarjAM, http://maarjam.botany.ut.ee/). This database enabled the description of 282 "virtual taxa" based on phylogenetic analyses of all collated DNA sequences (Öpik et al. 2010). As we will discuss later, this approach in combination with novel high-throughput metagenomic technology can lead to a "reverse taxonomy", thereby significantly alleviating the challenges imposed by ecologically important but rather inconspicuous microorganisms.

Another example of the difficult task of recognizing small sized biota can be portrayed by the investigation of benthic meiofauna (45–500 μm sized metazoa). Recent findings using a high throughput parallel sequencing approach (see below) on eight samples from an intertidal estuarine beach revealed unprecedented levels of diversity, including highly abundant groups for which little information is currently available (i.e. Platyhelminthes) (Fonseca et al. 2010). Moreover, non-asymptotic taxa accumulation curves were achieved for all of the detected phyla, suggesting that current estimations of species diversity may be fundamentally flawed.

Animal studies have demonstrated how DNA barcoding can be used to approach biodiversity surveys of understudied taxa and areas. Smith et al. (2009) estimated accumulation curves of the diversity of parasitoid wasps from Churchill, Manitoba, using both morphological and COI DNA barcoding phylogenetic diversity analyses. Parasitic wasps lack in morphological diagnostic characters, and the target area located near the Canadian arctic is relatively understudied. Contrary to the accumulation curves based on morphological data, those derived from DNA barcodes did not reach an asymptote. The same pattern was observed using an additional nuclear locus (i.e. the large subunit (LSU) rRNA gene region). Smith et al. (2009) concluded that these differences likely reflect undescribed species diversity among Churchill's parasitic wasps. They also propose that for certain understudied taxonomic groups and regions, the proposed approach could be the most immediate solution to address pressing questions regarding biodiversity assessment, conservation and estimation of sampling efficiency, while formal species boundaries await resolution under an integrative taxonomy framework.

In the process of developing BOLI, important compromises were made to enable the use of different DNA barcoding regions for major organism lineages other than animals. Since the concept is based on using small DNA sequence fragments (i.e. barcodes) that enable fast identification by allowing a sufficiently large "DNA barcoding gap", novel non-COI based barcoding regions have been proposed for fungi (Seifert et al. 2007; Seifert 2009; Begerow et al. 2010; Bellemain et al. 2010) and plants (Chase et al. 2005; Kress et al. 2005; Newmaster et al. 2006; Kress and Erickson 2007) in an attempt to reach consensus for universal DNA barcodes. In the case of fungi, although the historical use by mycologists of nuclear ribosomal internal transcribed spacer (ITS) sequences for species differentiation lead to its general acceptance as marker (Seifert 2009), this is not problem free (Bellemain et al. 2010), and important groups, such as the widespread AMF require a nuclear ribosomal DNA sequence of *ca.* 1,500 bp, stretching across the small subunit (SSU), ITS and part of the LSU regions for species level discrimination (Stockinger et al. 2010). For plants, although there have been proposals for universal barcodes, including those with capacity to reveal cryptic species in certain groups (e.g. Orchidaceae) (Lahaye et al. 2008), the CBOL Plant Working Group recently proposed a 2-loci combination of plastid coding regions (*rbcL* and *matK*) as the barcode for land plants (Hollingsworth et al. 2009). Compared with animals, species level resolution has proved difficult to achieve for plants. Nevertheless, the new proposed barcode has an approximate 70% capacity of species discrimination.

"Armed" with these markers, biologists now have the opportunity to merge their efforts by populating global repositories. Although the discussed deviations from the original concept and approach (i.e. one barcode region for all species) impose further logistical challenges for the generalized application of DNA barcoding in the discovery and monitoring of biodiversity, this is somehow counterbalanced by technical innovations such as the emergence of the high-throughput second generation sequencing (see below).

There are abundant examples of the utility of DNA barcoding for species discovery or traceability. In addition to the ones described above, the ability to identify pieces from bits of tissue, to assign eggs and larvae to species, or identify prey in stomach contents, are among the most relevant strengths of this approach. Below we discuss and describe applications of DNA barcoding in the framework of biodiversity discovery and monitoring, highlighting foreseeable subsidiary gains to our understanding of ecosystem functioning.

4.4.1 Cryptic Species

Cryptic species complexes are groups of distinct, reproductively isolated species with similar morphology (Saez and Lozano 2005). They are particularly intriguing from an evolutionary standpoint and represent a significant portion of unaccounted biodiversity (Pfenninger and Schwenk 2007). Taxonomists have been aware of the existence of cryptic species even before the system of classification developed by Linnaeus was widely adopted (Winker 2005). However, without DNA sequencing capability to rapidly analyse many individuals within species and across their distribution ranges, the task of characterizing patterns of cryptic species abundance and distributions is extremely limited. Adaptation in these species may have occurred for various and important traits, however, not for those morphologically evident. For instance, adaptations to novel biotic interactions or abiotic factors such as climate change may constitute a valuable biological resource and, thus, a priority for conservation (see Bickford et al. 2007). The comprehensive diagnosis of cryptic species across all taxa may reveal random and potential non-random trends, which would contribute to a better understanding of evolutionary biology and bio-geographical patterns of biodiversity.

The contribution of DNA barcoding to the discovery of new species without compulsory need for morphological descriptions is perhaps less prone to criticism for cryptic species due to the putative lack of appropriate traits for those descriptions in the first place. For example, using DNA barcodes, Smith et al. (2006) demonstrated the existence of host-specific species of parasitoid flies previously thought to be generalists (i.e. polyphagous). Their findings suggest that cryptic species of parasitoid flies are much more abundant than previously thought. In addition, these findings have important implications in areas such as invasive species biology and management (see below). DNA barcoding can be a valuable tool to speed up responses to biological invasions by helping in the discovery of invasive cryptic species or of species that can be tested as putative biological control agents. While the work by Smith et al. (2006) focused on species with a narrow distribution range, others have revealed that previously considered single species with a large distribution range were indeed several species with seemingly morphologically identical adults but different juveniles with preference for different resources (Hebert et al. 2004a). Given that only species with overlapping distribution ranges were considered, one can only imagine the number of cryptic species across the entire distribution range of this species complex.

Since the publication of the paper by Hebert et al. (2003a), DNA barcodes have been contributing to taxonomic clarifications in a wide array of organisms (Teletchea 2010). Among the many different published studies, a sizeable proportion report putative or confirmed cryptic species. At least for animals there is consolidated evidence for cryptic species in a diverse set of taxonomic groups, such as bryozoans (Gomez et al. 2007) amphipods (Witt et al. 2006) and wasps (Smith et al. 2008). Some authors argue that cryptic species are homogeneously distributed across taxa (Pfenninger and Schwenk 2007), in which case the relevance of molecular approaches as DNA barcoding to augment taxonomic resolution and accuracy of species discovery is evident. Indeed, even among vertebrate taxa, there have been studies suggesting unexpectedly high levels of cryptic diversity. In a DNA barcoding survey of fish species common to Australia and South African waters Zemlak et al. (2009) found sufficient COI-barcode divergence among fish populations to suggest the presence of overlooked species. This pattern was common in 8 out of 20 shore dwelling species, but absent in species restricted to offshore areas. This led the authors to propose that a complete survey of nearly 1,000 fish species shared between these two regions, may reveal as many as 300 hidden species.

As we have seen, even in a reasonably well-studied taxonomic assemblage such as fish, DNA barcoding may have a significant contribution to reveal hidden diversity and patterns of genetic differentiation associated with ecological attributes. If DNA barcoding can provide such exceptional insights to diversity in the case of a well-known group of organisms, one may hypothesize how much progress could be achieved in less traceable taxa, particularly with the assistance of emerging technologies as discussed in the next subsection.

4.4.2 Second-Generation Sequencing and the Biodiversity of Complex Environmental Samples

Since the discovery of the polymerase chain reaction (PCR) (Mullis et al. 1987) for DNA amplification, approximately 15 years before the concept of DNA barcoding was proposed, that species identification in environmental samples using DNA-PCR based techniques follows the same standard approach: (1) DNA extraction, (2) amplification of a marker of interest from a "target" group, (3) confirmation of amplification by gel-electrophoresis, (4) cloning of PCR products in *Escherichia coli* plasmids and (5) sequencing. To minimize steps four and five, many "fingerprinting" methods, that enable the discrimination of similar sized DNA sequences based on their different nucleotide arrangements, have been developed (Kowalchuk et al. 2004). Nevertheless, putative taxa identification by these methods is only possible through contrasts against reference DNA sequence libraries. Generally, all DNA "fingerprinting" methods are expensive, time consuming and typically fail to detect taxa present in low abundance. In addition, determination of relative species abundance is by and large inadequate and, it is not uncommon for available primers to have specificity issues and bias amplification towards certain taxonomic groups (e.g. Avis et al. 2006; Anderson et al. 2003). Although these problems persist in the novel approaches discussed hereafter, a major contrast is the sheer number of sequences retrieved, which may indeed contribute to alleviate them.

Over the past decade the method of pyrosequencing, which is based on the principle of "sequencing by synthesis", is emerging as the main method of next-generation sequencing (NGS) (i.e. after Sanger sequencing, the technology that has been used for 30 years; Sanger et al. 1977). NGS consists of high-throughput sequencing platforms that enable sequencing millions of DNA fragments in parallel (see Hall 2007; Voelkerding et al. 2009). The company 454 Life Sciences further developed pyrosequencing into an increasingly popular platform for large-scale DNA sequencing in parallel. 454 sequencing, as it became known (to our knowledge the numbers have no scientific meaning), combines pyrosequencing with emulsion PCR (http://www.454.com), allowing greater simplification (e.g. no need to clone DNA sequences into *E. coli* due to ultra-high throughput by simultaneous emulsion PCR—a form of *in vitro* cloning—and nucleotide reads), lower prices and sequence reads of suitable length for DNA barcoding. Roche Diagnostics currently owns the technology and sells a 454 benchtop sequencer that allows sequencing 400–600 million bp per run with 400–500 bp sequence lengths (soon to go up to 1,000 bp). Although 454 sequencing can be used for several applications, being especially practical to assemble whole genomes, it is now the preferred choice for species identification in complex samples (i.e. metagenomics). Basically, after DNA amplification from environmental samples, products from each sample can be tagged and, after emulsion PCR, placed into a "PicoTiterPlate" that allows reading simultaneously approximately 1 million DNA sequences per run.

The combination of DNA barcoding with NGS allows the study of biodiversity in ways that were impossible to afford just 5 years ago. In addition to detecting rare species in complex samples more effectively than traditional cloning coupled with Sanger sequencing, it can provide reliable measures of relative abundance (i.e. taxon rarefaction curves) due to the very large quantity of DNA sequences being simultaneously processed. This allows distinguishing sequences that, due to their high abundance, are likely to belong to ecologically important organisms whose identity is anonymous. As such, the approach can prompt further isolation attempts and proper description of organisms carrying those sequences; an approach named 'reverse taxonomy' as proposed by Markmann and Tautz (2005). Although the number of studies combining barcoding and NGS are still modest, we predict that it will increase substantially in the next decade.

DNA barcoding coupled with NGS is a powerful technique with large potential to further elucidate the biodiversity of environments that, in spite of their notably high species richness and complexity still remain largely unexplored by science. Examples of complex environments are soils, aquatic sediments or even the water column of certain nutrient rich ecosystems where plankton is abundant. There is also a very practical need to swiftly identify species in these environments. Benthic macroinvertebrate communities, for example, are important components of aquatic ecosystems, being routinely used in the assessment of aquatic ecosystem health (Metcalfe 1989). Under the European Union's Water Framework Directive, member states must implement an aquatic ecosystem-monitoring network by 2015. However, to achieve this goal there is a requirement for novel approaches that significantly speed-up benthic macroinvertebrate monitoring, which is traditionally a

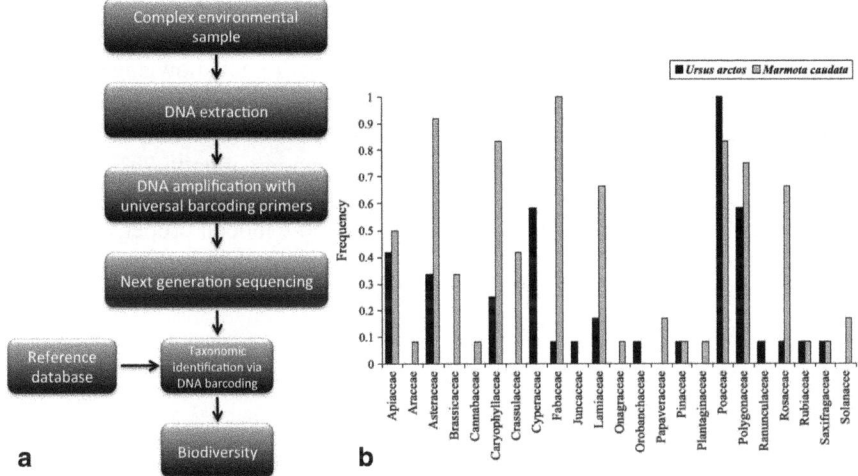

Fig. 4.2 **a** Diagram showing the steps of DNA barcoding coupled with next-generation sequening (NGS) for determination of biodiversity composition in complex environmental samples, and **b** Empirical data comparing diet compositions obtained from NGS analysis of faeces of the golden marmot (*Marmota caudata*) and brown bear (*Ursus arctos*) in Deosai National Park (Pakistan). (Based on Valentini et al. (2009a) with permission from John Wiley and Sons. The authors provide species level data in a separate table)

time-consuming undertaking (Baird and Sweeney 2011). Barcoding coupled with NGS may help resolve this need as indicated by recent work on mesofauna (Fonseca et al. 2010).

Using the ITS region of ribosomal DNA, Buée et al. (2009) assessed fungal diversity in different forest soils using 454 sequencing. The overwhelming amount of data acquired ("no less than 166 350 ITS reads") is in sharp contrast with the number of matches with known species and, as the authors point out, this "calls for curated sequence databases". The same limitations are patent in other NGS fungal metagenomic studies, namely of AMF (e.g. Öpik et al. 2009). Despite these current drawbacks, NGS of soil samples consistently reveals a much larger diversity than anticipated, and suggests that the combination of barcoding with NGS can serve as a catalyst for the discovery of new species and the understanding of spatio-temporal dynamics of soil biodiversity.

DNA barcoding coupled with NGS can help to elucidate and resolve important questions in ecology. For example, it can clarify the structure of food webs, resource partitioning between competing species or aid in conservation plans for species at risk. This can be achieved by using DNA barcoding-NGS to analyse consumer diets (i.e. samples of stomach contents or faeces) (Valentini et al. 2009a, b) (Fig. 4.2). Comparative tests in healthy and post-mortem specimens can even elucidate possible causes for increased mortality in certain species. Despite the novelty of these approaches several studies demonstrate their potential. Soininen et al. (2009) investigated the diets of sub-arctic small herbivores, comparing DNA-barcoding/

NGS with traditional histological methods. The former approach proved to be more reliable, revealing "more detailed results". Deagle et al. (2009) determined what animals eat through DNA barcoding/NGS from feaces. To our knowledge, there are no DNA barcoding/NGS studies on plant communities. However, if one imagines that entire experimental plots can now be harvested, the shoot material ground-up, mixed and NGS ran with a level of species discrimination of up to 70% as well as having the capacity to capture their relative abundances, one can only conclude that science was never closer to unraveling the spatio-temporal properties of plant communities.

4.4.3   Invasive Species

Most countries are now interconnected through the globalized economy where products circulate as dictated by private sector capital and the general public's demand. The expedite circulation of millions of goods by air, land or sea represents an unprecedented number of pathways, providing numerous possibilities for the introduction of species far beyond their native range. In absence of human activity these alien species would not be able to reach and establish in those new territories. As the number of markets with which each country trades increases, so does the number of alien species introductions (Lin et al. 2007). Although only a very small number of species establish and become naturalized in their introduced range and, of these, an even smaller number becomes invasive (i.e. their growth and spread becomes high enough to cause environmental damage and economic loss), one single invasive species is able to cause major disruptions to ecosystems (Simberloff and Rejmánek 2011). Indeed, invasive species have been considered a major cause of species extinctions (Wilson 1992). Despite this view, it is rather difficult to establish direct causal effects between invasive species and species extinctions (Gurevitch and Padilla 2004). In part, this is because only a very small portion of the biota have been taxonomically described and catalogued. Furthermore, it is logistically challenging, time-consuming and costly to monitor species diversity in invaded areas. By contributing to the upgrade of taxonomy (see above), DNA barcoding can play a key role not only in the early detection of introduced species but also in the monitoring of native biota responses to the invasive species. Despite the novelty of the concept, several examples of ongoing initiatives and studies demonstrate its promise.

Many potentially invasive species arrive at border crossings at incipient or cryptic life-cycle stages, often impossible to identify by traditional means. The Quarantine Barcode of Life Initiative (QBOL), aims to develop a "new diagnostic tool using DNA barcoding to identify quarantine organisms in support of plant health". This initiative (for details see www.qbol.org) requires extensive international research networking with linkages to policy for the development of up-to-date lists of quarantine organisms. Even though the initiative should reduce the need for centralized diagnostic laboratories with many staff specialized in different taxa, it can contribute to different facets of the modern taxonomy as described earlier (e.g. integration of barcoding and morphological data into databases, discovery of novel potentially invasive species through "reverse taxonomy"). QBOL is currently focused on the European continent. An International Network for Barcoding Invasive and Pest Species (INBIPS, http://www.barcoding.si.edu/INBIPS.htm) was created at the first First International Barcode Conference in London with the objective to facilitate barcoding activities on invasive species issues. The more individuals and organizations join these initiatives the more DNA barcodes and taxonomic features will become available, thereby making databases like QBOL powerful tools to quickly screen traded products. In turn, this will prevent the introduction of invasive species, including those present in complex products such as seed blends.

Chown et al. (2008) used DNA barcoding to determine that immature specimens of a moth species present in the sub-Antarctic were indeed exotic. Juvenile forms of insect pests are sometimes impossible to tell apart from benign species using morphological traits. Moreover, identification of adults may require intricate dissections by experts. Recent DNA barcoding research enabled the identification of different species of insect pests (i.e. leafminers) (Scheffer et al. 2006). This is particularly useful for prevention and management of agricultural pests because false positives can be problematic, significantly affect trade and, thus, cause unnecessary and costly responses. Although Scheffer et al. (2006) had to use other published data, including nuclear genes, to validate their results, DNA barcoding contributed to elucidate the taxonomy of leafminers, demonstrating that a universal database approach in a "wiki" base style may be the most cost-effective approach (see above). As more data, including morphological and other DNA sequences, are gathered for economically important groups, DNA barcoding will become increasingly effective as a tool for rapid detection of invasive species.

By being an affordable and standardized method, DNA barcoding enables the collection of massive amounts of comparable data globally, across different laboratories. This information can be particularly useful, especially when spiked with data from additional genes, for phylogeographical comparisons aimed at investigating population level genetic patterns of invasive species between their native and introduced ranges (Yassin et al. 2008). In addition to providing clues about dispersal history, these comparisons can indicate what source population(s) founded the exotic one(s), being potentially advantageous in the quest for biological control agents. On a wider scale they can help elucidate why some species become invasive, although this may require combining additional trait information. Finally, as described earlier, DNA barcoding can help discover cryptic species in the native range, some carrying traits associated with invasibility while others do not. In the near future, coupling DNA barcoding with NGS may help improve the effectiveness of this approach. Moreover, it will contribute to understand whether certain structural features of native communities (i.e. species richness and their relative abundances) determine their susceptibility to biological invasions. This information would enable detecting areas at risk, thereby allowing the implementation of preventive measures.

Although most efforts have focused on animals, DNA barcoding has also contributed to improvements in the taxonomy of invasive plant species. The genus *Acacia*, for example, includes many cryptic species that are aggressive invaders in different parts of the world (Maslin et al. 2003). Differentiation among invasive species from morphologically identical but rare species is relevant from a conservation standpoint. Using DNA barcoding, Newmaster and Ragupathy (2009) were able to discriminate cryptic species of *Acacia* and to differentiate biogeographical patterns among populations present in different continents.

As the scientific community accepts more universal barcoding primers for different taxa, it is likely that DNA barcoding will expand to play a key role in the prevention, early detection and control of invasive species as well as in the expedite assessment of their impact on native ecosystems.

4.5 Conclusion

Towards the end of the twentieth century it became apparent that the extant biodiversity was much greater than the operational infrastructure and human resources available to study it. Because of the taxonomic impediment, other pure and applied disciplines dependent on data downstream of alpha taxonomy would also experience constraints to develop a more comprehensive understanding of natural biological systems.

The emergence of new technologies, with special relevance to informatics and genomics, offers new hope to tackle the impediment and meet the challenges of biodiversity discovery and indexing. There are abundant indications that the science of taxonomy already initiated a process of deep transformation through incorporation of new technologies in its practice, and by concomitant organizational re-shaping. These changes are expected to benefit the output of taxonomic science, which will meet more effectively the demand for accurate knowledge of biological diversity from other disciplines and end-users. By meeting these needs, taxonomy may gain the much-deserved broader and sustained support from policy-makers and society in general, ultimately re-gaining a more prominent status among the biological sciences.

The barcode of life initiative and the DNA barcoding approach hold particular features that make them important catalysts of these changes in taxonomy, and prominent players in the new operational paradigm. The future will tell how DNA barcoding practice will evolve within the broader transformation in course. Will it become standard practice to add molecular information to morphological species descriptions, as for example in the form of one or multiple DNA barcodes? Will it be possible to DNA barcode comprehensively museum collections and type specimens? What system will be implemented by the scientific community to curate the newly built massive databases, specimen collections and tissue banks? As new technologies become available and the sequencing costs decrease will DNA barcoding evolve into a multi-locus DNA barcode? Embedded in an advanced twenty-first century taxonomy, and used extensively in biodiversity monitoring, DNA barcoding (and closely related approaches) may contribute to generate ground-breaking insights into spatial-temporal dynamics of species, in a multitude of communities and ecosystems, many of them still poorly known.

Some skeptics forecast that DNA barcoding will induce taxonomic chaos and the consequent wholesale subtraction of scientific rigor of taxonomy and its professionals. However, a higher confidence in the judgment capability of science and scientists would be expected: if DNA barcoding is as harmful for science as it is claimed, then it will be readily dismissed by the scientific community as soon as its flaws emerge, which so far has not been the case. Moreover, DNA barcoding cannot thrive in isolation. Its progress is contingent upon concurrent and harmonized development of the whole scientific endeavor of discovery and knowledge of life's diversity. For this purpose, skilled taxonomists will be sought after and, if anything, DNA barcoding initiatives are just increasing the demand for expert input (e.g. Packer et al. 2009). Regardless of what the future brings for DNA barcoding, the transformation of taxonomy is already in motion, and "Man's obsessive quest to catalogue life" (see Dunn 2009) will endure; from this point onwards it can only get sharper.

Acknowledgments This is a contribution from F.O. Costa in the scope of grants PTDC/MAR/69892/2006 from "Fundação para a Ciência e a Tecnologia", and a European Commission's Reintegration Grant (ERG-224890). P.M. Antunes thanks the Natural Sciences and Engineering Research Council of Canada (NSERC) and the Ontario Ministry of Natural Resources (OMNR) for financial support. We thank the Consortium for the Barcode of Life for permission to reproduce Fig. 4.1, and John Wiley and Sons publishers for granting permission to reproduce Fig. 4.2. Special thanks to Luisa Borges and Ana Cunha for comments on an early draft of the manuscript.

References

Agnarsson I, Kuntner M (2007) Taxonomy in a changing world: seeking solutions for a science in crisis. Syst Biol 56:531–539

Anderson IC, Campbell CD, Prosser JI (2003) Potential bias of fungal 18S rDNA and internal transcribed spacer polymerase chain reaction primers for estimating fungal biodiversity in soil. Environ Microbiol 5:36–47

Antunes PM, Koch AM, Morton JB, Rillig MC, Klironomos JN (2011) Evidence for functional divergence in arbuscular mycorrhizal fungi from contrasting climatic origins. New Phytol 189:507–514

Appeltans W et al. (eds) (2010) World register of marine species. http://www.marinespecies.org. Accessed 03 Feb 2011

Avis PG, Dickie IA, Mueller GM (2006) A 'dirty' business: testing the limitations of terminal restriction fragment length polymorphism (TRFLP) analysis of soil fungi. Mol Ecol 15:873–882

Baird DJ, Sweeney BW (2011) Applying DNA barcoding in benthology: the state of the science. J N Am Benthol Soc 30:122–124

Begerow D, Nilsson H, Unterseher M, Maier W (2010) Current state and perspectives of fungal DNA barcoding and rapid identification procedures. Appl Microbiol Biotechnol 87:99–108

Bellemain E, Carlsen T, Brochmann C, Coissac E, Taberlet P, Kauserud H (2010) ITS as an environmental DNA barcode for fungi: an in silico approach reveals potential PCR biases. BMC Microbiol 10:189

Besansky NJ, Severson DW, Ferdig MT (2003) DNA barcoding of parasites and invertebrate disease vectors: what you don't know can hurt you. Trends Parasitol 19:545–546

Bickford D et al. (2007) Cryptic species as a window on diversity and conservation. Trends Ecol Evol 22:148–155

Bidartondo MI et al. (2008) Preserving accuracy in GenBank. Science 319:1616

Blaxter ML (2004) The promise of a DNA taxonomy. Philos Trans R Soc Lond B Biol Sci 359:669–679

Boero F (2010) The study of species in the era of biodiversity: a tale of stupidity. Diversity 2:115–126

Bouchet F (2006) The magnitude of marine biodiversity. In: Duarte CM (ed) The exploration of marine biodiversity: scientific and technological challenges. Fundación BBVA, Bilbao, pp 31–62

Bucklin A et al. (2010) DNA barcoding of Arctic ocean holozooplankton for species identification and recognition. Deep Sea Res Pt II 57:40–48

Bucklin A, Steinke D, Blanco-Bercial L (2011) DNA barcoding of marine metazoa. Ann Rev Mar Sci 3:18.11–18.38

Buée M et al. (2009) 454 Pyrosequencing analyses of forest soils reveal an unexpectedly high fungal diversity. New Phytol 184:449–456

Buhay JE (2009) "COI-Like" sequences are becoming problematic in molecular systematic and DNA barcoding studies. J Crustacean Biol 29:96–110

Carvalho GR (1998) Molecular ecology: origins and approach. In: Carvalho GR (ed) Advances in molecular ecology. IOS Press, Amsterdam

Carvalho MR de et al. (2007) Taxonomic impediment or impediment to taxonomy? A commentary on systematics and the cybertaxonomic-automation paradigm. Evol Biol 34:140–143

Chapman AD (2009) Numbers of living species in Australia and the world. Report for the Australian biological resources study. Department of the Environment, Water, Heritage and Arts of the Australian Government, Canberra

Chase MW et al. (2005) Land plants and DNA barcodes: short-term and long-term goals. Philos Trans R Soc B Biol Sci 360:1889–1895

Chown S, Sinclair B, Vuuren B van (2008) DNA barcoding and the documentation of alien species establishment on sub-Antarctic Marion Island. Polar Biol 31:651–655

Costa FO, Carvalho GR (2007a) The barcode of life initiative: reply to Dupré, Hollingsworth and Holm. Genomics Soc Policy 3:52–56

Costa FO, Carvalho GR (2007b) The barcode of life initiative: synopsis and prospective societal impacts of DNA barcoding of fish. Genom Soc Policy 3:29–40

Costa FO, Carvalho GR (2010) New insights into molecular evolution: prospects from the Barcode of Life Initiative (BOLI). Theory Biosci 129:149–157

Costa FO et al. (2007) Biological identifications through DNA barcodes: the case of the Crustacea. Can J Fish Aquat Sci 64:272–295

Dasmahapatra KK, Mallet J (2006) DNA barcodes: recent successes and future prospects. Heredity 97:254–255

Dayrat B (2005) Towards integrative taxonomy. Biol J Linn Soc 85:407–415

De Queiroz K (2007) Species concepts and species delimitation. Syst Biol 56:879–886

Deagle BE, Kirkwood R, Jarman SN (2009) Analysis of Australian fur seal diet by pyrosequencing prey DNA in faeces. Mol Ecol 18:2022–2038

DeSalle R (2006) Species discovery versus species identification in DNA barcoding efforts: response to Rubinoff. Conserv Biol 20:1545–1547

Dunn R (2009) Every living thing: man's obsessive quest to catalog life, from nanobacteria to new monkeys. HarperCollins, NY

Dupré J (2007) Real but modest gains from genetic barcoding. Genomics Soc Policy 3:41–43

Ebach MC, Carvalho MR de (2010) Anti-intellectualism in the DNA barcoding enterprise. Zoologia 27:165–178

Ebach MC, Holdrege C (2005) DNA barcoding is no substitute for taxonomy. Nature 434:697

Ellis R, Waterton C, Wynne B (2010) Taxonomy, biodiversity and their publics in twenty-first-century DNA barcoding. Public Underst Sci 19:497–512

Evenhuis NL (2007) Helping solve the "other" taxonomic impediment: completing the eight steps to total enlightenment and taxonomic nirvana. Zootaxa 1407:3–12

Fitter AH (2005) Darkness visible: reflections on underground ecology. J Ecol 93:231–243

Floyd R, Abebe E, Papert A, Blaxter M (2002) Molecular barcodes for soil nematode identification. Mol Ecol 11:839–850

Fonseca VG et al. (2010) Second-generation environmental sequencing unmasks marine metazoan biodiversity. Nat Commun 1:98

Froese R, Pauly D (2010) FishBase. World Wide Web electronic publication. http://www.fishbase.org. Accessed Nov 2010

Gamper HA, Walker C, Schussler A (2009) Diversispora celata sp nov: molecular ecology and phylotaxonomy of an inconspicuous arbuscular mycorrhizal fungus. New Phytol 182:495–506

Gershoni M, Templeton AR, Mishmar D (2009) Mitochondrial bioenergetics as a major motive force of speciation. Bioessays 31:642–650

Godfray HCJ (2002) Challenges for taxonomy—the discipline will have to reinvent itself if it is to survive and flourish. Nature 417:17–19

Godfray HCJ (2006) To boldly sequence. Trends Ecol Evol 21:603–604

Godfray HCJ, Knapp S (2004) Taxonomy for the twenty-first century—introduction. Philos Trans R Soc Lond B Biol Sci 359:559–569

Goldstein PZ, DeSalle R (2010) Integrating DNA barcode data and taxonomic practice: determination, discovery, and description. Bioessays 33:135–147

Gomez A, Wright PJ, Lunt DH, Cancino JM, Carvalho GR, Hughes RN (2007) Mating trials validate the use of DNA barcoding to reveal cryptic speciation of a marine bryozoan taxon. Proc R Soc B Biol Sci 274:199–207

Gottelli N, Colwell RK (2001) Quantifying biodiversity: procedures and pitfalls in the measurement and comparison of species richness. Ecol Lett 4:379–391

Gurevitch J, Padilla DK (2004) Are invasive species a major cause of extinctions? Trends Ecol Evol 19:470–474

Hajibabaei M, Janzen DH, Burns JM, Hallwachs W, Hebert PDN (2006) DNA barcodes distinguish species of tropical Lepidoptera. Proc Natl Acad Sci U S A 103:968–971

Hajibabaei M, Singer GAC, Hebert PDN, Hickey DA (2007) DNA barcoding: how it complements taxonomy, molecular phylogenetics and population genetics. Trends Genet 23:167–172

Hall N (2007) Advanced sequencing technologies and their wider impact in microbiology. J Exp Biol 210:1518–1525

Hanner R (2005) Data standards for BARCODE records in INSDC (BRIs). http://barcodingsiedu/PDF/DWG_data_standards-Finalpdf

Harris DJ (2003) Can you bank on GenBank? Trends Ecol Evol 18:317–319

Hebert PDN, Gregory TR (2005) The promise of DNA barcoding for taxonomy. Syst Biol 54(5):852–859

Hebert PDN, Cywinska A, Ball SL, DeWaard JR (2003a) Biological identifications through DNA barcodes. Proc R Soc Lond B Biol 270:313–321

Hebert PDN, Ratnasingham S, deWaard JR (2003b) Barcoding animal life: cytochrome c oxidase subunit 1 divergences among closely related species. Proc R Soc Lond B Biol 270:S96–S99

Hebert PDN, Penton EH, Burns JM, Janzen DH, Hallwachs W (2004a) Ten species in one: DNA barcoding reveals cryptic species in the neotropical skipper butterfly Astraptes fulgerator. Proc Natl Acad Sci U S A 101:14812–14817

Hebert PDN, Stoeckle MY, Zemlak TS, Francis CM (2004b) Identification of birds through DNA barcodes. PLoS Biol 2:1657–1663

Hey J (2006) On the failure of modern species concepts. Trends Ecol Evol 21:447–450

Hollingsworth PM (2007) DNA barcoding: potential users. Genomics Soc Policy 3:44–47

Hollingsworth PM et al. (2009) A DNA barcode for land plants. Proc Natl Acad Sci U S A 106:12794–12797

Holm P (2007) The book of life goes online. Genom Soc Policy 3:48–51

Janzen DH (2004) Now is the time. Philos Trans R Soc Lond B Biol Sci 359:731–732

Jaume D, Duarte CM (2006) General aspects concerning marine and terrestrial biodiversity. In: Duarte CM (ed) The exploration of marine biodiversity: scientific and technological challenges. Fundación BBVA, Bilbao, pp 17–30

Johnson GD et al. (2009) Deep-sea mystery solved: astonishing larval transformations and extreme sexual dimorphism unite three fish families. Biol Lett 5:235–239

Knapp S, Polaszek A, Watson M (2007) Spreading the word. Nature 446:261–262

Koch AM, Croll D, Sanders IR (2006) Genetic variability in a population of arbuscular mycorrhizal fungi causes variation in plant growth. Ecol Lett 9:103–110

Kowalchuk GA, Bruijn FJd, Head IM, Akkermans AD, Elsas JD van (2004) Molecular microbial ecology manual, 2nd edn. Kluwer Academic, Dordrecht

Kress WJ, Erickson DL (2007) A two-locus global DNA barcode for land plants: the coding *rbcL* gene complements the non-coding *trnH-psbA* spacer region. PLoS One 2:e508

Kress WJ, Wurdack KJ, Zimmer EA, Weigt LA, Janzen DH (2005) Use of DNA barcodes to identify flowering plants. Proc Natl Acad Sci U S A 102:8369–8374

Lahaye R et al. (2008) DNA barcoding the floras of biodiversity hotspots. Proc Natl Acad Sci U S A 105:2923–2928

Lane N (2009) On the origin of bar codes. Nature 462:272–274

Lin W, Zhou G, Cheng X, Xu R (2007) Fast economic development accelerates biological invasions in China. PLoS One 2:e1208

Lyal CHC, Weitzman AL (2004) Taxonomy: exploring the impediment. Science 305:1106

Markmann M, Tautz D (2005) Reverse taxonomy: an approach towards determining the diversity of meiobenthic organisms based on ribosomal RNA signature sequences. Philos Trans R Soc B Biol Sci 360:1917–1924

Maslin BR, Miller JT, Seigler DS (2003) Overview of the generic status of Acacia (Leguminosae: Mimosoideae). Aust Syst Bot 16:1–18

Metcalfe JL (1989) Biological water quality assessment of running waters based on macroinvertebrate communities: history and present status in Europe. Environ Pollut 60:101–139

Moritz C (2002) Building the biodiversity commons. D-Lib Magazine 8. http://wwwdliborg/dlib/june02/moritz/06moritzhtml

Moritz C, Cicero C (2004) DNA barcoding: promise and pitfalls. PLoS Biol 2:1529–1531

Mullis KB, Erlich HA, Arnheim N, Horn GT, Saiki RK, Scharf SJ (1987) Process for amplifying, detecting, and/or-cloning nucleic acid sequences. US 4683195 United StatesThu Feb 07 16:07:56 EST 2008 Patent and Trademark Office, Box 9, Washington, DC 20232.NOV; NOV-87-074136; EDB-87-172157 English

Newmaster SG, Ragupathy S (2009) Testing plant barcoding in a sister species complex of pantropical *Acacia* (Mimosoideae, Fabaceae). Mol Ecol Resour 9:172–180

Newmaster SG, Fazekas AJ, Ragupathy S (2006) DNA barcoding in land plants: evaluation of rbcL in a multigene tiered approach. Can J Bot Rev Can Bot 84:335–341

Öpik M, Metsis M, Daniell TJ, Zobel M, Moora M (2009) Large-scale parallel 454 sequencing reveals host ecological group specificity of arbuscular mycorrhizal fungi in a boreonemoral forest. New Phytol 184:424–437

Öpik M et al. (2010) The online database MaarjAM reveals global and ecosystemic distribution patterns in arbuscular mycorrhizal fungi (Glomeromycota). New Phytol 188:223–241

Packer L, Grixti JC, Roughley RE, Hanner R (2009) The status of taxonomy in Canada and the impact of DNA barcoding. Can J Zool 87:1097–1110

Padial JM, De La Riva I (2007) Integrative taxonomists should use and produce DNA barcodes. Zootaxa 1586:67–68

Padial JM, De La Riva I (2010) A response to recent proposals for integrative taxonomy. Biol J Linn Soc 101:747–756

Padial JM, Miralles A, De la Riva I, Vences M (2010) The integrative future of taxonomy. Front Zool 7:16

Patterson DJ, Cooper J, Kirk PM, Pyle RL, Remse DP (2010) Names are key to the big new biology. Trends Ecol Evol 25:689–691

Pawlowska TE, Taylor JW (2004) Organization of genetic variation in individuals of arbuscular mycorrhizal fungi. Nature 427:733–737

Pfenninger M, Schwenk K (2007) Cryptic animal species are homogeneously distributed among taxa and biogeographical regions. BMC Evol Biol 7:121

Pleijel F et al. (2008) Phylogenies without roots? A plea for the use of vouchers in molecular phylogenetic studies. Mol Phylogenet Evol 48:369–371

Polaszek A et al. (2005) A universal register for animal names. Nature 437:477

Raven PH (2004) Taxonomy: where are we now? Philos Trans R Soc Lond B Biol Sci 359:729–730

Rodman JE, Cody JH (2003) The taxonomic impediment overcome: NSF's partnerships for enhancing expertise in taxonomy (PEET) as a model. Syst Biol 52:428–435

Rubinoff D (2006) DNA barcoding evolves into the familiar. Conserv Biol 20:1548–1549

Saez AG, Lozano E (2005) Body doubles. Nature 433:111

Sanger F, Nicklen S, Coulson AR (1977) DNA sequencing with chain-terminating inhibitors. Proc Natl Acad Sci U S A 74:5463–5467

Savolainen V, Cowan RS, Vogler AP, Roderick GK, Lane R (2005) Towards writing the encyclopaedia of life: an introduction to DNA barcoding. Philos Trans R Soc B 360:1805–1811

Schander C, Willassen E (2005) What can biological barcoding do for marine biology? Mar Biol Res 1:79–83

Scheffer S, Lewis ML, Joshi RC (2006) DNA barcoding applied to invasive leafminers (Diptera: Agromyzidae) in the Philippines. Ann Entomol Soc Am 99:204–210

Schindel DE (2010) Biology without borders. Nature 467:779–781

Schindel DE, Miller SE (2005) DNA barcoding a useful tool for taxonomists. Nature 435:17

Schlick-Steiner BC, Steiner FM, Seifert B, Stauffer C, Christian E, Crozier RH (2010) Integrative taxonomy: a multisource approach to exploring biodiversity. Annu Rev Entomol 55:421–438

Seifert KA (2009) Progress towards DNA barcoding of fungi. Mol Ecol Resour 9:83–89

Seifert KA et al. (2007) Prospects for fungus identification using C01 DNA barcodes, with Penicillium as a test case. Proc Natl Acad Sci U S A 104:3901–3906

Sieverding E, Oehl F (2006) Revision of Entrophospora, and description of Kuklospora and Intraspora, two new genera in the arbuscular mycorrhizal Glomeromycetes. J Appl Bot Food Qual Angew Bot 80:69–81

Simberloff D, Rejmánek M (2011) Encyclopedia of biological invasions. University of California Press, Los Angeles

Smith VS (2005) DNA barcoding: perspectives from a "Partnerships for Enhancing Expertise in Taxonomy" (PEET) debate. Syst Biol 54:841–844

Smith SE, Read DJ (2008) Mycorrhizal symbioses. Academic Press, London

Smith MA, Woodley NE, Janzen DH, Hallwachs W, Hebert PDN (2006) DNA barcodes reveal cryptic host-specificity within the presumed polyphagous members of a genus of parasitoid flies (Diptera: Tachinidae). Proc Natl Acad Sci U S A 103(10):3657–3662

Smith MA et al. (2008) Extreme diversity of tropical parasitoid wasps exposed by iterative integration of natural history, DNA barcoding, morphology and collections. Proc Natl Acad Sci U S A 105:12359–12364

Smith MA, Fernandez-Triana J, Roughley R, Hebert PDN (2009) DNA barcode accumulation curves for understudied taxa and areas. Mol Ecol Resour 9:208–216

Soininen E et al. (2009) Analysing diet of small herbivores: the efficiency of DNA barcoding coupled with high-throughput pyrosequencing for deciphering the composition of complex plant mixtures. Front Zool 6:16

Stockinger H, Kruger M, Schussler A (2010) DNA barcoding of arbuscular mycorrhizal fungi. New Phytol 187:461–474

Tautz D, Arctander P, Minelli A, Thomas RH, Vogler AP (2003) A plea for DNA taxonomy. Trends Ecol Evol 18:70–74

Teletchea F (2010) After 7 years and 1000 citations: comparative assessment of the DNA barcoding and the DNA taxonomy proposals for taxonomists and non-taxonomists. Mitochondr DNA 21:206–226

Valentini A et al. (2009a) New perspectives in diet analysis based on DNA barcoding and parallel pyrosequencing: the trnL approach. Mol Ecol Resour 9:51–60

Valentini A, Pompanon F, Taberlet P (2009b) DNA barcoding for ecologists. Trends Ecol Evol 24:110–117

Voelkerding KV, Dames SA, Durtschi JD (2009) Next-generation sequencing: from basic research to diagnostics. Clin Chem 55:641–658

Walters C, Hanner R (2006) Platforms for DNA banking. In: De Vicente MC, Andersson MS (eds) DNA Banks—providing novel options for Gene Banks? Topical reviews in agricultural biodiversity. International Plant Genetic Resources Institute, Rome, pp 22–35

Ward RD, Hanner R, Hebert PDN (2009) The Campaign to DNA barcode all fishes, FISH-BOL. J Fish Biol 74:329–356

Ward RD, Zemlak TS, Innes BH, Last PR, Hebert PDN (2005) DNA barcoding Australia's fish species. Philos Trans R Soc B 360:1847–1857

Wheeler QD (2004) Taxonomic triage and the poverty of phylogeny. Philos Trans R Soc Lond B Biol Sci 359:571–583

Wheeler QD (2008) The new taxonomy. Systematics association special volumes. CRC Press, Boca Raton

Wheeler QD, Raven PH, Wilson EO (2004) Taxonomy: impediment or expedient? Science 303:285

Will KW, Mishler BD, Wheeler QD (2005) The perils of DNA barcoding and the need for integrative taxonomy. Syst Biol 54:844–851

Wilson EO (1992) The diversity of life. Questions of science. Belknap Press of Harvard University Press, Cambridge

Wilson EO (2003) The encyclopedia of life. Trends Ecol Evol 18:77–80

Wilson EO (2004) Taxonomy as a fundamental discipline. Philos Trans R Soc Lond B Biol Sci 359:739

Winker K (2005) Sibling species were first recognized by William Derham (1718). Auk 122:706–707

Witt JDS, Threloff DL, Hebert PDN (2006) DNA barcoding reveals extraordinary cryptic diversity in an amphipod genus: implications for desert spring conservation. Mol Ecol 15:3073–3082

Yassin A, Capy P, Madi-Ravazzi L, Ogereau D, David JR (2008) DNA barcode discovers two cryptic species and two geographical radiations in the invasive drosophilid Zaprionus indianus. Mol Ecol Resour 8:491–501

Zemlak TS, Ward RD, Connell AD, Holmes BH, Hebert PDN (2009) DNA barcoding reveals overlooked marine fishes. Mol Ecol Resour 9:237–242

Chapter 5
Increasing Pressure at the Bottom of the Ocean

Ricardo Serrão Santos, Telmo Morato and Fernando J. A. S. Barriga

Abstract Invisibly hidden under the waters, the deep sea has been considered to be the least affected habitat on Earth by human use. However, recently, the perception of the damage and its extent are coming to light.

The ocean is recognisably under threat due to a number of direct human activities, of which fishing industry and pollution are of major concern. Other emergent economic activities such as mining, the extraction of oil and gas, and the sequestration of CO_2, should be evaluated beforehand to take into account the forecasting and mitigation of possible impacts.

These human activities migrated to the deep-sea, fisheries and waste deposit first, followed by oil and mineral exploitation. This is reflected in the growing number of species and habitats requiring conservation actions and the need for new management instruments for the deep ocean. In particular one has to take into consideration that the majority of these habitats and associated species are located on the high seas where the capacity for intervention and the legal basis either do not exist or may fall far short of what is needed (Probert et al., Seamounts: ecology, fisheries and conservation, 2007).

5.1 Introduction

Deep, dark, wide and susceptible to remarkable pressure, the deep ocean is the largest component of the surface of our planet and therefore the most unknown (Ramirez-Llodra et al. 2010). Satellites, the new technologies of observation introduced in the twentieth century, are unable to scrutinize the oceans with the same efficiency as for terrestrial habitats and ecosystems. The oceans are rather opaque. Visual observation is only possible with cameras attached to robots and through submersibles operated

R. S. Santos (✉) · T. Morato
Department of Oceanography and Fisheries at the University of Azores, IMAR—Institute of Marine Research and LARSyS Associated Laboratory, 9901-862 Horta (Azores), Portugal
e-mail: ricardo@uac.pt

F. J. A. S. Barriga
Department of Geology, Faculty of Sciences, University of Lisbon, CREMINER-FCUL and LARSyS Associated Laboratory, 1749-016 Lisbon, Portugal

A. Mendonca et al. (eds.), *Natural Resources, Sustainability and Humanity*, 69
DOI 10.1007/978-94-007-1321-5_5, © Springer Science+Business Media Dordrecht 2012

by large oceanic vessels, an extremely expensive operation. For that reason only a small portion of the deep ocean has been scientifically studied properly.

Much of this vast "ecosystem" consists of abyssal plains covered with sediments, usually characterized by low productivity (and dependent on organic matter falling from superficial zones), low physical energy, low macrobiological diversity and low biological rates. Other important components of the deep sea are the solid prominent topographic rocky structures, such as the mid-ocean ridges, seamounts and the hydrothermal vent fields.

"Ridges" continuously lacerate all oceans of our planet. Seamounts, knolls and hydrothermal massif fields are discreetly distributed in space, sometimes organized in clusters (for definitions see Pitcher et al. 2007). In contrast to the abyssal plains, habitats composing these areas are usually very productive, with biomasses of up to 20 kg/m^2 in the hydrothermal sources, compared to about 1 g/m^2 that can be found on soft bottoms of the abyssal plains (Martins et al. 2008; Wei et al. 2010).

5.2   The Importance of the Oceans

Huge 70% of the planet's surface and more than 90% of the available volume: 170 times more space available to life than all other combined ecosystems.

Unknown In deep sea only an area corresponding to a few football fields has actually been sampled from a scientific point of view.

Diverse From the thirty-six known phyla, 35 occur in marine environments where 14 of them are endemic, compared with only 1 in terrestrial environments. A recent study by Mora et al. (2009) suggest that the predicted number of marine species is 2.2 million, of which 91% still waiting description. Most of these to be discovered in the deep-sea.

Miscellaneous deep ocean is a miscellany with several levels of productivity and biomass. For example in the abyssal plains we generally found low biomass but high biodiversity, particularly small organisms and bacteria that live buried in its soft bottoms, whereas in hydrothermal ecosystems we found very high biomass but low biological diversity with high levels of endemic species.

Rich In the seafloor and sub-seafloor there are abundant metal deposits. The deep sea is home of hundreds of million tones of polymetallic massive sulphide deposits, containing, in several proportions, metals such as gold, silver, copper and zinc. Others, including cobalt, are abundant in crusts and nodules.

5.3   Deepwater Fisheries

Sustainability of living marine resources is one of the most problematic components concerning deep sea management. The fishing industry is in a remarkable crisis since more than 50% of the fishing stocks are operating at their maximum limit, while around 30% are overexploited, depleted or recovering (FAO 2010, Fig. 5.1).

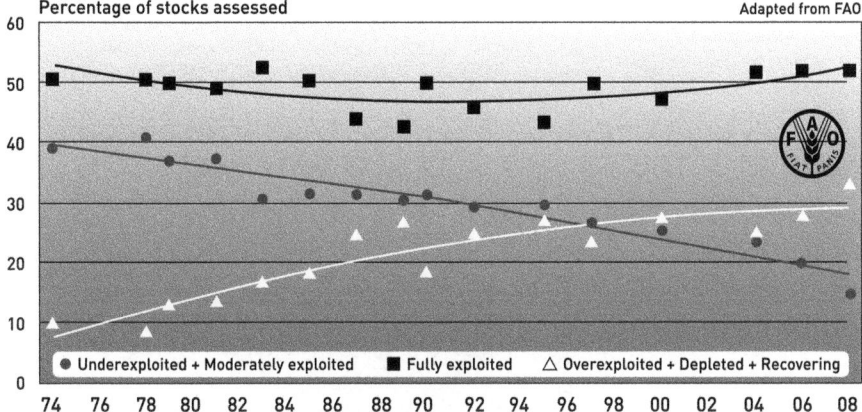

Fig. 5.1 Global trends in the state of world marine stocks from 1974 until 2008. According to FAO the proportion of stocks estimated to be underexploited or moderately exploited declined from 40% in the mid-1970s to 15% in 2008. In contrast, the proportion of overexploited, depleted or recovering stocks increased from 10% in 1974 to 32% in 2008. The proportion of fully exploited stocks has remained relatively stable at about 50% since the 1970s. (© Figure and information taken from FAO 2010)

The extended crisis of fisheries on traditional fishing grounds on continental shelves led to the increased exploitation of deeper water resources (Fig. 5.2) and to the discovery of fish aggregations at seamounts to where fishing fleets migrated in the 1980s and 1990s. This extended the fishing impacts to deeper waters and highs seas (Morato et al. 2006b). A relevant part of this fleet was/is composed of deep-sea trawlers, which quickly over-exploited the new stocks of fish such as the orange roughy and the oreos in New Zealand and South Australia in the 1980s, and subsequently in the North Atlantic (Clark et al. 2007). Exploitation of seamounts quickly spread throughout the oceans. In all cases there was a rapid depletion of living resources (Fig. 5.2).

On the other hand, intensive trawling in the deep-sea eventually produce incidental catch of benthic organisms (i.e. that are stuck or associated with the bottom of sea). These fisheries lead to habitat degradation with effects on the local biodiversity and biomass of the benthic species. Some of the relevant impacted species are deep-water corals and sponges, whose longevity can reach thousands of years (Marina Carreiro Silva et al. in preparation), as it is the case of *Leiopathes* sp. in the Azores (Fig. 5.3).

Deep-water fisheries, particularly those using trawls and gill nets are the most pressing and harmful current threats to the deep-sea. However the actual drilling for extraction of hydrocarbons, which reaches depths of 3,000 m at the seafloor level and around 7,000 m on sub-seafloor, and marine seafloor mining, may be also posing significant threats (see next section). All these activities can affect especially highly vulnerable deep-water corals and sponges that form organic reefs (Hall-Spencer et al. 2002; Roberts 2002; Roberts et al. 2006). The recovery of these

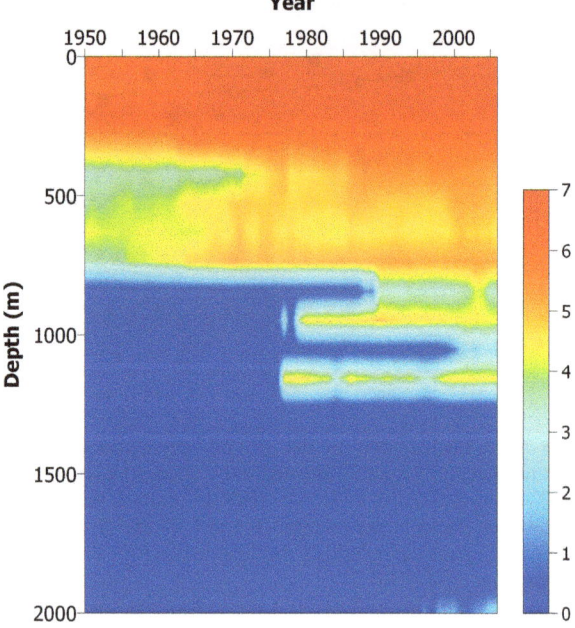

Fig. 5.2 Time series of world marine bottom fisheries catches by depth strata, between 1950 and 2006, showing the increased exploitation of the deep ocean. *Coloured scale* show catch in tonnes log10 transformed

environments can take hundreds to thousands of years, but it is thought that in some cases they never recover (Roberts et al. 2006). They are not comparable to anything we know about coastal and shallow habitats. Deep water corals and sponges are rather important as structuring habitats, contributing to increasing biodiversity and biomass (OSPAR 2010).

The question that arises is whether it is possible a sustainable management of deep-sea living resources? Fisheries moved, in recent years, to deep-sea species (Morato et al. 2006b). It is known that deep water fish have a high intrinsic vulnerability (Morato et al. 2006a) due to a number of unique features of its life cycle, such as slow growth, high longevity (e.g. the orange roughy can reach 150 years old), late maturity (e.g. at 40 years old also in the orange roughy), low natural mortality, etc. These factors make deep-sea fishes more sensitive to exploitation. It is generally agreed that deep trawl and gill nets are equivalent to "mining" and may lead to the extirpation of commercial species. Deep-sea living resources sustainability seems to be possible only where traditional, small-scale fishing practice exists, as in some archipelagic areas (Silva and Pinho 2007) where deep-sea trawls are banned, as it is the case in the Azores (Probert et al. 1997).

In terms of research, it is required to compare the population dynamics of fish stocks where fishing with traditional methods still exists (i.e. with hook and line and deep-sea "long-line" from vessels until 30 m) and equivalent species are exploited through industrial methods such as bottom trawls and automated "long-line". A recent study held in the Azores shows that small scale fisheries is the most sustainable activity with better results in terms of stocks resilience, habitat integrity and socio-

Fig. 5.3 Gorgonian (or *black coral*) of the genus *Leiopathes* photographed in August 2010 on Condor seamount during a mission of the EU-FP7 project CoralFish (NRP "Gago Coutinho" with the ROV Luso of EMEPC)

economic value (Morato and Pitcher 2005; Carvalho et al. 2011). But even in these regions the need of protecting entire seamounts is advisable (Santos et al. 1995, 2009, 2010, Fig. 5.4).

Industrial deep-sea fishing is considered as non-renewable (Glover and Smith 2003). Apart from the direct effects on target species and fish and invertebrate by-catch, mostly wasted, also causes extensive and excessive habit disruption (Gianni 2004). The consequences need to be carefully evaluated and studied. From the perspective of future research, comprehensive studies based on monitoring the affected areas will be necessary, as well as the observation of pristine areas and the establishment of protected area in the high seas. On the other hand, it will be required small-scale experimental studies involving the removal of organisms to follow up the process of colonization succession in time and space. A recent implementation of a scientific observatory on the Condor seamount at the Azores (Morato et al. 2010) is expected to provide clues on these questions.

The question that arises is not whether if we have already affected the deep-sea biotic resources, but if they can be recovered? Some deep-sea habitats, such as the abyssal plains with sediment substrates, when disturbed by human activity such as trawl fishing and mining may eventually recover in a relatively short period of time. Others such as seamounts, where filter feeding organisms such as corals and sponges form aggregations, can take hundreds to thousands of years if they come to be recovered.

But fisheries are not the only human activity migrating to the deep-sea: oil and gas exploitation, mineral extraction, and CO_2 sequestration, are other actual or forthcoming examples.

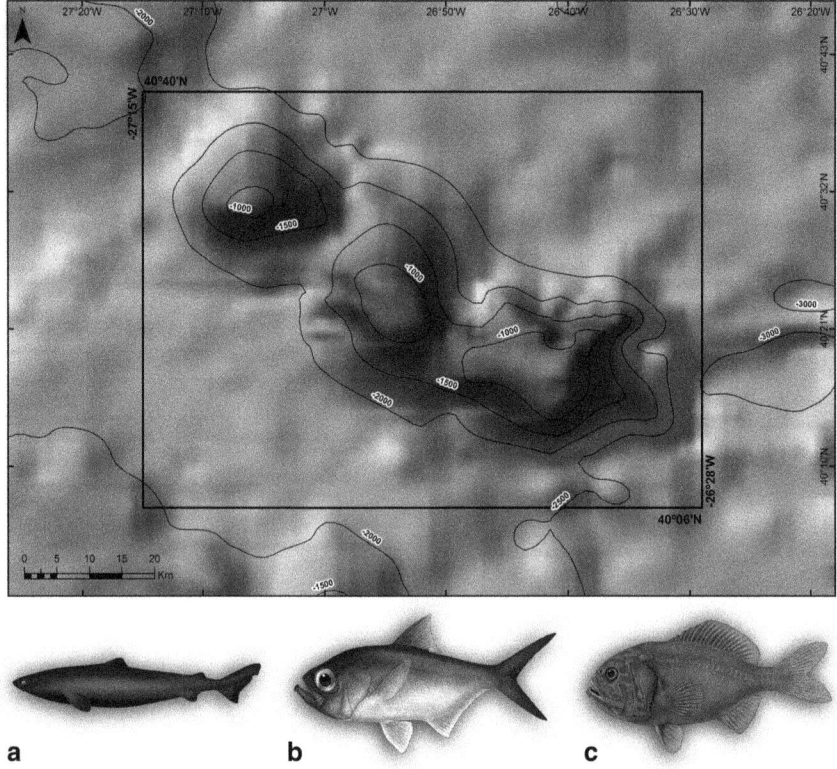

Fig. 5.4 Marine protected area of Sedlo Seamount (F Tempera & R Medeiros © *Imag*DOP). Three fish species that occur there, which are priority species in terms of conservation: **a** *Centroscymnus coelolepis*—Portuguese dogfish, **b** *Beryx splendens*—alfonsino, **c** *Hoplostethus atlanticus*—orange roughy. (Illustrations by Les Gallagher © fishpics & *Imag*DOP)

5.4   Oil, Gas and Minerals Extraction

The extraction of oil and gas from the marine sub-seafloor is an important, complex and long established economic and commercial activity. For these reasons it has nowadays well-established rules that aim to assess and minimize the negative impacts over environment and ecosystems in the exploitation zone of influence (Glover and Smith 2003). But the recent case of the "Deepwater Horizon oil spill" in the Gulf of Mexico caused extended damage in habitats, fishing grounds and even in the tourism industry. The impacts were widespread to the deep seafloor, the water column and coastal area. The company involved had great difficulties to control the spill which lasted for 3 months, starting at around 9,900 m^3 of crude oil per day. The effects of the oil spill will persist for years, but there are already clear signs of recovery.

In contrast to oil extraction, mineral exploitation in the deep-sea is still in an initial phase. The industry is using technologies adapted from the offshore oil and gas industry (http://www.nautilusminerals.com). Nautilus Minerals is the international company acting in Papua New Guinea. After some years of exploration they got concessions for commercial exploitation of Seafloor Massive Sulphide systems, containing high grade copper, gold, zinc and silver.

In the near future, it is expected a growing interest in deep-sea mining for polymetallic sulphide deposits. The Russian Federation submitted recently an application to the International Seabed Authority with of an exploration plan for polymetallic sulphides in the Area, in the Mid-Atlantic Ridge (www.isa.org.jm/en/node/627) between 10° and 16°N. China also applied and it is expected that other countries will follow, including Germany.

5.5 Manganese Nodules and Crusts

One of the potential mineral exploitation activities often mentioned as the more economic interesting is related to manganese nodules. These nodules are found at great depths on the marine seafloor. All nodules contain a nucleus composed of a core solid material that can be a fishbone, shark tooth, sea-urchin quill or fragments of volcanic rock. The manganese crusts and nodules are among the mineral deposits of slower formation: from 1 to 20 mm in a million years. The crust thickness ranges from less than 1 up to 240 mm (commonly between 20 and 40 mm thick) (Verlaan 1992). The nodules are polymetallic compounds which include nickel, cobalt, manganese and copper, but also gold, silver, titanium and zinc. The amounts of nickel, cobalt and manganese in nodules oceanic deposits exceed the reserves in the continents.

The impact of mining activities on deep-sea benthic life, as the exploitation of manganese nodules, has been the subject of several studies (e.g. Ahnert and Schriever 2001; Borowski 2001; Vopel and Thiel 2001) as a result of hundreds of cruise explorations led by international consortia. However, none of these studies thoroughly assessed the impact of the plumes produced during the extraction process that spread through the water column. There are no reliable impact studies concerning the exploitation of manganese crusts and polymetallic deposits. While nodules are a surface layer of loose potato-sized concretions, polymetallic sulphide deposits are largely loose, but thicker, and may exist sub-seafloor; crusts stick firmly to the underlying seabed.

5.6 Methane Hydrates

Another deep ocean potential resource are methane hydrates. Methane hydrates have the appearance of ice and occur at the bottoms of several seas and oceans. These derived from methane molecules released during the degradation of organic matter by bacteria, or from inorganic sources, such as serpentinization or mantle degassing. Methane is trapped in water crystals that eventually form hydrates. The crystals of methane hydrate can be distributed over several hundred meters below the marine seafloor. Besides its great potential as a primary fuel, methane hydrates as a whole contain unique bacterial habitats. It is known that 60% of bacteria living on Earth are in sediments in marine sub-seafloor. It is estimated that the volume of gas in the world reservoirs of methane hydrates exceeds the volume of the total fossil fuel reserves in the continents.

However, the exploitation and use of methane poses a lot of threats. Methane, as a green house effect gas, has direct effects in global warming.

5.7   Polymetallic Sulphides of the Hydrothermal Fields

The polymetallic sulphide deposits occur in back-arc and mid-ocean spreading ridges. Chimneys are only a small part of the mineral deposits that are formed at or near the seafloor by seawater that penetrates into the sub-seafloor, through pores and fissures, is heated and reacts with the rocks, extracting several metals, rising again, extremely hot, with or without a component of magmatic fluid. The typical hydrothermal vent chimneys are simply deposits of the minerals formed from precipitation of the components dissolved in modified sea-water during this process (Fig. 5.5). These chimneys of metal deposits contain copper, zinc, gold, silver and other metals. At the same time, these sites house chemosynthetic communities of great biological and, potentially, biotechnological interest.

5.8   Potential Impacts of Mineral and Gas Extraction

Certainly, mineral extraction will grow in importance in the future and will have commercial interest (Barriga and Santos 2010). The oil and gas at deeper levels in the sub-seafloor will also have great demand in the near future. New technological solutions are expected for exploitation (Wiltshire 2001). Therefore more studies on both benthic and pelagic environments and impact over organisms and habitats are required.

There is 3,000 times more methane in the deep sea than anywhere else on Earth. The extraction is inevitable as a substitute for conventional fossil fuels. But methane has significant implications in climate. The main environmental concerns related to its exploitation will be focused on this aspect. Besides, methane extraction will have effects over the microbiological and bacterial communities living in these environments. It is unknown and difficult to foresee the relevance and extent of this impact.

Deep-sea mineral extraction uses the technology of self-propelled machines with vibrators which crush the crusts for hydraulic lift from depths of 800–3,000 m (Chung 2005). When manganese nodules are collected, the fine sediment also extracted is spread over the surface proliferating like plumes in the water column and

Fig. 5.5 Deep sea vent chimney at the Rainbow hydrothermal field (−2,300 m deep). (Photo: © Mission SEAHMA (FCT-PDCTM/P/MAR/15281/1999 w/IFREMER))

thus potentially affecting large areas of the oceans. The process also leads to direct disturbance in benthic organisms at the operation site.

5.9   The Deep Sea as a Deposit

The oceans, particularly the deep sea, were once considered to have infinite capacity to dilute and recycle any substances, making them harmless. For this reason the deep sea has become, in the twentieth century, the elected place for the disposal of an array of substances, e.g. radioactive, heavy metals, carbon, and others (Tyler 2003). These threats to marine life and also to human kind were already identified, leading to significant changes in waste disposal procedures. The storage of radioactive substances in the deep sea has now practically stopped. Currently, the biggest challenge arises from the concept of carbon sequestration under the deep-seafloor.

The deposit/sequestration of CO_2, produced by industrial activities, in oceanic geological formations is being considered for some years. The plan seems inevitable and experiments occur in this direction. The solution may raise some problems or side effects such as the induction of changes in pH and changes in local biochemistry, which can cause related changes in local biological communities.

However, given the vastness of the deep sea, its seafloor is in fact less affected than equivalent shallow areas. It is expected that many countries will require the removal of CO_2 through sea bottom sequestration, namely in deep sea sediments with remineralisation. The impacts on local wildlife should not be neglected (Glover and Smith 2003).

Few available studies (Grassle and Morse-Porteous 1987; Shirayama 1997; Watanabe et al. 2006; Keil et al. 2010) present disparate conclusions. While deep sea organisms may react differently from their counterparts of coastal areas, also some natural deep sea processes may be analogues of potential "contaminated" disposals (Tyler 2003), it has also been proved that sequestered CO_2 can negatively affect deep-sea sediment-dwelling animals (Thistle et al. 2005).

5.10   Priority Habitats and Ecosystems

Several conventions, committees, councils and directives (e.g. OSPAR, ICES, IUCN, CBD, NATURA 2000, etc.) have recently defined a set of habitats and species that require urgent management actions, e.g. the seamounts, deep-sea coral reefs and gardens, sponges' aggregations, oceanic ridges with hydrothermal systems, etc. There is also a growing interest in the creation of Marine Protected Areas (Gjerde and Breide 2003). These processes are a result of the recognition of real and current threats to resources, biodiversity and the deep-sea habitats.

5.11   Final Considerations

Fisheries are one of the best documented human activities. The depressing effect regarding target species, but also non-target species and habitats (or collateral effects) does not raise any doubts any more that have reached the deep sea (Morato et al. 2006b). Deep-water fisheries are considered equivalent to "mining living resources" (Glover and Smith 2003). It is very likely that some of deep-sea fishing resources will no longer recover from the effects of over-exploitation, while the damages to habitats caused by deep-sea trawling may also be irreversible all over the place where they so far occurred.

Regarding mineral exploitation activities we may say that they still are at the beginning or in a conceptual stage. However, no matter the strain these activities might exert in the system, they will be unavoidable, given the foreseen exhaustion of traditional mineral deposits. Oil exploitation in marine environment is already a reality for several years and technology has ensured developments that provide exploitation at increasingly deeper levels. The environmental management for exploitation and the control of environmental impacts, although improved over the years, had some setbacks (the most recent and dramatic being the "Deepwater Horizon" oil spill in the Gulf of Mexico). The environment negative impacts that can be anticipated related to mineral exploration in ocean depths are difficult to foretell based on the experiments so far conducted.

Concerning contamination through waste disposal, including radioactive substances and heavy metals storage is known for a long time and there are international efforts to control them. The effects of CO_2 sequestration need to be evaluated and anticipated.

The international scientific community and international governmental and nongovernmental organizations are increasingly engaged in the debate concerning the criteria for exploiting the bottom of the sea, especially in areas beyond national jurisdiction where the legal framework is ambiguous (Halfar and Fujita 2002; Young 2003; Gjerde 2006).

As the future belongs to mankind, solutions should be brought forward to ensure the sustainability and balance of the planet and its oceans, while cautioning that the actions to be developed in the future will be based on rigorous scientific knowledge of the impact of economic activities. The scientific community at large must be called to participate, identifying stresses on the marine ecosystems, monitoring economic activities and conducting experiments. There is an on-going international process in view the implementation of a network of Marine Protected Areas in the Deep Sea and Open Ocean based on scientific criteria established under the umbrella of the Convention of Biological Diversity (CBD 2009).

Acknowledgments RSS acknowledges Ângela Mendonca (chair of the 2nd International School Congress on "Natural Resources, Sustainability and Humanity") for invitation to participate and present two lectures at the 2nd International School Congress in Braga and Maria José Marques and Ana Cunha for editorial comments and improvements. Iva Flores helped with revision of the English.

We acknowledge the Portuguese Foundation for Science and Technology (FCT-Lisbon) and the Regional Azorean Directorate for Science, Technology and Communications (DRCTC-Azores), for funding of IMAR/DOP/UAz Research Unit #531, CREMINER FCUL and the Associated Laboratory LARSyS.

This paper contributes to HERMIONE (grant agreement no. 226354) and to CoralFish (grant agreement no. 213144) projects funded by the European Community's Seventh Framework Programme (FP7/2007–2013).

References

Ahnert A, Schriever G (2001) Response of abyssal Copepoda Harpacticoida (Crustacea) and other meiobenthos to an artificial disturbance and its bearing on future mining for polymetallic nodules. Deep Sea Res II 48:3779–3794

Barriga FJAS, Santos RS (2010) Recursos minerais marinhos, metálicos, não metálicos e energéticos: potencial e impactos ambientais. In: Vieira CMN, Soromenho MV, Falcato J, Leitão AG (eds) Políticas Públicas do Mar. Esfera do Caos, Lisboa, pp 86–95, 297

Borowski C (2001) Physically disturbed deep-sea macrofauna in the Peru Basin, southeast Pacific, revisited 7 years after the experimental impact. Deep Sea Res II 55:55–81

Carvalho N, Edwards-Jones G, Isidro E (2011) Defining scale in fisheries: small versus large-scale fishing operations in the Azores. Fish Res 109:360–369

CBD (2009) Azores scientific criteria and guidance for identifying ecologically or biologically significant marine areas and designing representative networks of marine protected areas in open ocean waters and deep sea habitats. Secretariat of the Convention on Biological Diversity, Montréal, p 12

Clark MR, Vinnichenko VI, Gordon JDM, Beck-Bulat GZ, Kukharev NN, Kakora AF (2007) Large-scale distant-water trawl fisheries on seamounts (XXIV). In: Pitcher TJ, Morato T, Hart PJB, Clark MR, Haggan N, Santos RS (eds) Seamounts: ecology, fisheries and conservation. Blackwell, Oxford, pp 361–399

Chung JS (2005) Deep-ocean mining technology: development II. Proceedings of The Sixth ISOPE Ocean Mining Symposium, Chansgha, Hunan, China

FAO (2010) The state of world fisheries and aquaculture 2010. FAO, Rome, p 197

Gianni M (2004) High seas bottom trawl fisheries and their impacts on the biodiversity of vulnerable deep-sea ecosystems. IUCN, Gland, p 90

Gjerde K (2006) Ecosystems and biodiversity in deep waters and high seas. UNEP Reg Seas Rep Stud 178:58

Gjerde KM, Breide C (2003) Towards a strategy for high seas marine protected areas: proceedings of the IUCN, WCPA and WWF experts workshop on high seas marine protected areas, 15–17 Jan 2003, Malaga, Spain. IUCN, Gland, Switzerland

Glover AG, Smith CR (2003) The deep-sea floor ecosystem: current status and prospects of anthropogenic change by the year 2025. Environ Conserv 30(3):219–241

Grassle JF, Morse-Porteous LS (1987) Macrofaunal colonization of disturbed deep-sea environments and the structure of deep-sea benthic communities. Deep Sea Res II 34:1911–1950

Halfar J, Fujita JM (2002) Precautionary management of deep-sea mining. Mar Policy 26:103–106

Hall-Spencer J, Allain V, Fossa JH (2002) Trawling damage to Northeast Atlantic ancient coral reefs. Proc R Soc Lond B Biol Sci 269(1490):507–511

Keil RG, Nuwer JM, Strand SE (2010) Burial of agricultural by products in the deep sea as a form of carbon sequestration: a preliminary experiment. Mar Chem 122:91–95

Martins I, Colaço A, Dando PR, Martins I, Desbruyères D, Sarradin PM, Marques JC, Santos RS (2008) Size-dependent variations on the nutritional pathway of Bathymodiolus azoricus demonstrated by a C-flux model. Ecol Model 217(1–2):59–71

Mora C, Tittensor DP, Adl S, Simpson AGB, Worm B (2009) How many species are there on earth and in the Ocean? PLoS Biol 9(8):e1001127

Morato T, Pitcher TJ (2005) Ecosystem simulations in support of management of data-limited seamount fisheries. In: Kruse GH, Gallucci VF, Hay DE, Perry RI, Peterman RM, Shirley TC, Spencer PD, Wilson B, Woodby D (eds) Fisheries assessment and management in data limited situations. Alaska Sea Grant, University of Alaska Fairbanks, Lowell Wakefield Fisheries Symposium Series 21, pp 467–486

Morato T, Cheung WWL, Pitcher TJ (2006a) Vulnerability of seamount fish to fishing: fuzzy analysis of life-history attributes. J Fish Biol 68(1):209–221

Morato T, Watson R, Pitcher TJ, Pauly D (2006b) Fishing down the deep. Fish Fish 7(1):24–34

Morato T, Pitcher TJ, Clark MC, Menezes G, Tempera F, Porteiro F, Giacomello E, Santos RS (2010) Can we protect seamounts for research? Oceanography 3(1):190–199

OSPAR (2010) Background document for coral gardens. Biodivers Ser (Publication No. 486/2010):40

Pitcher TJ, Morato T, Hart PJB, Clark MR, Haggan N, Santos RS (eds) (2007) Seamounts: ecology, fisheries and conservation. Blackwell, Oxford, xxvi+p527 with 156 illustrations

Probert PK, Christiansen S, Gjerde KJ, Gubbay S, Santos RS (2007) Management and conservation of seamounts (Chapter 20). In: Pitcher T, Morato T, Hart PJB, Clark MR, Haggan N, Santos RS (eds) Seamounts: ecology, fisheries and conservation. Blackwell, Oxford

Ramirez-Llodra R, Brandt A, Danovaro R, De Mol B, Escobar E, German CR, Levin LA, Arbizu PM, Menot L, Buhl-Mortensen P, Narayanaswamy BE, Smith CR, Tittensor DP, Tyler PA, Vanreusel A, Vecchione M (2010) Deep, diverse and definitely different: unique attributes of the world's largest ecosystem. Biogeosciences 7(9):2851–2899

Roberts CM (2002) Deep impact: the rising toll of fishing in the deep sea. Trends Ecol Evol 17:242–245

Roberts JM, Wheeler AJ, Freiwald A (2006) Reefs of the deep: the biology and geology of cold-water coral ecosystems. Science 312:543–547

Santos RS, Hawkins S, Monteiro LR, Alves M, Isidro EJ (1995) Marine research, resources and conservation in the Azores. Aquat Conserv Mar Freshw Ecosyst 5(4):311–354

Santos RS, Christiansen S, Christiansen B, Gubbay S (2009) Towards the conservation and management of Sedlo seamount: a case study. Deep Sea Res Part II: Top Stud Oceanogr 56(25):2720–2730

Santos RS, Tempera F, Menezes G, Porteiro F, Morato T (2010) Mountains in the sea: Sedlo Seamount, Azores. Oceanography 23(1):148–149

Shirayama Y (1997) Biodiversity and biological impact of ocean disposal of carbon dioxide. Waste Manag 17(5–6):381–384

Silva HM, Pinho MR (2007) Small scale fishing on seamounts (Chapter 16). In: Pitcher T, Morato T, Hart PJB, Clark MR, Haggan N, Santos RS (eds) Seamounts: ecology, fisheries and conservation. Blackwell, Oxford

Thistle S, Carman KR, Sedlacek L, Brewer PG, Fleeger JW, Barry JP (2005) Deep-ocean, sediment-dwelling animals are sensitive to sequestered carbon dioxide. Mar Ecol Prog Ser 289:1–4

Tyler PA (2003) Disposal in the deep sea: analogue of nature or faux ami? Environ Conserv 30(1):26–39

Verlaan PA (1992) Benthic recruitment and manganese crust formation on seamounts. Mar Biol 113(1):171–174

Vopel K, Thiel H (2001) Abyssal nematode assemblages in physically disturbed and adajacent sites of the eastern equatorial Pacific. Deep Sea Res II 48:3795–3808

Watanabe Y, Yamaguchi A, Ishidai H, Harimoto T, Suzuki S, Sekido Y, Ikeda T, Shirayama Y, Mac Takashi M, Ohsumi T, Ishizaka J (2006) Lethality of increasing CO_2 levels on deep-sea copepods in the western North Pacific. J Oceanogr 62(2):185–196

Watson R, Morato T (2004) Exploitation patterns in seamount fisheries: a preliminary analysis. In: Morato T, Pauly D (eds) Seamounts: biodiversity and fisheries. Fisheries centre. Research Reports, vol 12, no 5. University of British Columbia, Vancouver, pp 61–66, 78

Wei C-L, Rowe GT, Escobar-Briones E, Boetius A, Soltwede T, Caley MJ, Soliman Y, Huettmann F, Qu F, Yu Z, Pitcher CR, Haedrich RL, Wicksten MK, Rex MA, Baguley JG, Sharma J, Danovaro T, MacDonald IR, Nunnally CC, Deming JW, Montagna P, Lévesque M, Weslawski JM, Wlodarska-Kowalczuk M, Ingole BS, Bett BJ, Billett DSM, Yool A, Bluhm BA, Iken K, Narayanaswamy BE (2010) Global patterns and predictions of seafloor biomass using random forests. PLoS One 5(12):e15323

Wiltshire J (2001) Future prospects for the marine minerals industry. Underwater 13:40–44

Young TR (2003) Developing a legal strategy for high seas marine protected areas. Workshop: High Seas Marine Protected Areas, Malaga, p 36

Chapter 6
Biological Invasions and Global Trade

An Urgent Issue

Dora Aguin-Pombo

Abstract Biological invasions are a large-scale phenomenon considered after habitat loss, the major threat to world biodiversity. In the last decades due to global trade and improvement of transport, humans and their goods have moved around the globe at an increasing rate. As an outcome of human activities, introductions of alien species have increased significantly resulting in a substantially increment in the number of pests caused by exotic organisms. The continuous expansion of invasive species is responsible for a significant impact on biodiversity and natural resources, industries, commerce and human health. Today the expansion and acceleration of biological invasions is an ecological problem at planetary scale equivalent to some of the well-known environmental issues such as global warming and rainforest destruction. Teacher's understandings of this complex topic and appropriate environmental-based science curricula are important steps to educate future citizens to be capable to limit further introduction of invasive species.

6.1 Introduction

Biological invasion is a common but ambiguous term that refers to the geographic range expansion of non-native species into new areas outside its native distributional range, where they arrived by human activity, either deliberate or accidental (but see Valéry et al. 2008). In their new home, facing different challenges and growth conditions, the non-native species also known as 'exotic' or 'alien' species, can acquire a competitive advantage over native species that allow them to spread rapidly and become dominant. When exotic species interfere positively or negatively in a community or ecosystem in which they disperse, are known as *invasive*. However, the impacts reported rarely focus on the effects that invasive species have on native ecosystems, instead most tend to emphasize the negative effects of alien species in human activities. Following this human-centered view, the effects

D. Aguin-Pombo (✉)
Universidade da Madeira, 9000-390 Funchal, Madeira, Portugal
e-mail: aguin@uma.pt

CIBIO, Centro de Investigação em Biodiversidade e Recursos Genéticos,
Universidade do Porto, 4485-601 Vairão, Portugal

A. Mendonca et al. (eds.), *Natural Resources, Sustainability and Humanity,*
DOI 10.1007/978-94-007-1321-5_6, © Springer Science+Business Media Dordrecht 2012

of non-indigenous species can be classified as positive or negative but, most widely known, are the negative effects caused to agriculture, forestry, and livestock or to human and animal health. Regardless whether the species is native or exotic, those organisms that caused trouble to man and or to their goods are commonly known as pests. In the last decades this definition of pest has been enlarged to include also microorganisms, plants, animals and fungi that threaten in addition to humans, native species and ecosystems. Yet, a general conformity on the definition of pest and other terms important to studies on invasion ecology is still missing, which makes often difficult an understanding among different sectors and individuals involved. The terms related to invasive alien species used in this chapter follow mainly those given by the Convention on Biological Diversity (CBD) (see Table 6.1).

Despite the ambiguous terminology, the impact of invasive species is large and known worldwide. As a result of global trade commercial exchanges of goods and persons have increased significantly in the last decade (Loope and Howarth 2002). Following commercial pathways many thousands of exotic species were able to overcome natural barriers and spread around the world (Ding et al. 2008). Human-mediated colonization of exotic species is very fast and extends over greater distances, resulting in an extensive modification of natural habitats. This large-scale phenomena represent today one of the major threats to European biodiversity (Hulme 2007).

Although humans have spread invasive species intentionally or unintentionally worldwide since Neolithic times (Webb 1985), most alien species have been introduced only after the improvement of shipping since about 1500 AD (di Castri et al. 1990). Both agriculture and animal husbandry, are good examples of the introduction of species into new habitats. From the sixteenth to the nineteenth century, the history of introduction of alien species is closely related to the European exploration and colonization of the planet (Crosby 1986). But the rates at which species have circumvented natural barriers and invade new continents increased dramatically after globalization. The main components of this occurrence are a growing volume of transported goods and persons, increasing efficiency and speed of transports, advancing technologies and trade agreements (Bright 1999).

The number and diversity of alien species transported by trade is huge. In the United States alone about 50,000 alien species have been introduced (Pimentel et al. 2005). The increase in exotic species introductions generally has lead to a considerably greater number of pests caused by exotic organisms. An analysis of the detection of regulated pest species for 29 European countries published by the European and Mediterranean Plant Protection Organization (EPPO) showed that, for the period comprised between 1995 and 2004, there was a total of 8,889 interceptions of non-indigenous pests, among which insects were largely dominant (75.9%) (Roques and Auger-Rozenberg 2006). However, since inspectors usually examine only a small percentage of all cargo entrances, these pests may account only for a small number of the goods exchanged (Work et al. 2005).

Biological invasions can provoke economic, environmental, or social impacts. The economic impacts of exotic species are enormous. In the case of the United States, the annual losses due to invasive species have been quantified between 125 to 150 billion $ (McNeely et al. 2001) but, for most of the countries, the losses due

Table 6.1 Terms used to describe biological invasions

Terms	Description
Biological invasions	Consists of a species' acquiring a competitive advantage following the disappearance of natural obstacles to its proliferation, which allows it to spread rapidly and to conquer novel areas within recipient ecosystems in which it becomes a dominant population (Valéry et al. 2008)
Casual species	An exotic that may flourish and even reproduce occasionally in an area, but which do not form self-replacing populations, and which rely on repeated introductions for their persistence (Richardson et al. 2000)
Exotic species	A species, subspecies, or lower taxon introduced (intentionally or accidentally) outside its normal past or present distribution as a result of the human activities or of his domestic animals; includes any part, gametes, seeds, eggs, or propagules of such species that might survive and subsequently reproduce. Synonyms: *alien, adventive, foreign, immigrant, introduced, imported, non-native, non-indigenous, transported* (adapted from CBD 2000)
Eradication	The extirpation of the entire population of an alien species in a managed area; eliminating the invasive alien species completely (CBD 2000)
Establishment	The process of a species in a new habitat successfully reproducing at a level sufficient to ensure continued survival without infusion of new genetic material from outside the system (CBD 2000)
Introduction	The movement, by human agency, of a species, subspecies or lower taxon (including any part, gametes, seeds, eggs, or propagule that might survive and subsequently reproduce) outside its natural range (past or present). This movement can be either within a country or between countries (IUCN 2000) and can be intentional (*intentional introduction*) or unintentionally (*unintentional introduction*)
Invasive exotic species	An alien species whose establishment and spread threaten ecosystems, habitats or species with economic or environmental harm (CBD 2000)
Native species	A species, subspecies, or lower taxon occurring living within its natural range (past or present), including the area which it can reach and occupy using its own legs, wings, wind/water-borne or other dispersal systems, even if it is seldom found there. Synonym: *indigenous species* (CBD 2000)
Naturalized species	Alien species that reproduce consistently (cf. casual alien species) and sustain populations over more than one life cycle without direct intervention by humans (or in spite of human intervention); they often reproduce freely, and do not necessarily invade natural, semi-natural or human made ecosystems (CBD 2000)
Pest	An ambiguous term generally used to describe any species that conflicts with human interests (economic, aesthetic or ecological)
Weeds	Plants (not necessarily *alien*) that grow in sites where they are not wanted and have detectable negative economic or environmental effects; alien weeds are invasive alien species. Synonyms: *plant pests, harmful species; problem plants* (CBD 2000)

to alien species have not been quantified yet. Conscious of the role of invasive species to society, economy and environment, the European Union through the European Commission published a Communication entitled *Halting the loss of biodiversity by 2010, and beyond—Sustaining ecosystem services for human wellbeing* which clearly highlights as a priority the need to diminish the effects of alien species. This document establishes that two of the main European objectives concerning biodiversity are to reach a significant decrease of the impact of invasive alien species and alien genotypes on EU biodiversity and to reduce the effects of international trade on both global biodiversity and ecosystem services (European Commission 2006).

Mooney and Drake (1989) raised several fundamental questions important to the initial stage of dispersal of biological invasions: who are the invaders, how do they get there, and where do they come from? Because humans are a key component of this phenomenon, if we intend to reduce new invasions we have to concentrate on the ways people facilitate the transport of species into new areas. This chapter reviews the vectors and pathways of exotic species related to global trade and their impacts to humans and ecosystems. To prevent further expansion of invasive species, the role of school teachers in developing science literacy and strengthening opinions and decisions about science-based issues is discussed.

6.2 Impacts Caused by Invasive Species

6.2.1 Impact of Invasive Species on Human Activities

The introduction and transport of exotic invasive species has often had large ecological disturbance and socioeconomic impacts both on aquatic and terrestrial ecosystems. Exotic species negatively affect agriculture, animal husbandry and other man use of natural resources besides of being also a major concern for human health. The worldwide losses due to invasions of arthropods caused to agriculture alone, ranges from 55 to 248 billion $ a year (Bright 1999). There are two different actions by which humans enhanced the impact of invasive species. First, we increased the natural dispersion of species either accidentally, through transportation of species (in soil, ballast, machinery) or, intentionally, by the importation of commodities for food such as animals, fruits or vegetables or plants for gardening and agriculture. Second, we have facilitated the establishment of certain species through the modification and disturbance of natural habitats.

The global impacts caused by invasive species particularly to agriculture are huge. Numerous exotic plants have the ability to reduce or make unprofitable crops and livestock farming (e.g. *Ulex europaeus*) (Norton 2009; Rees and Hill 2001). Some invasive plant species as the Spanish flag (*Lantana camara*) are poisonous, and can caused the death of cattle and sheep (McLennan and Amos 1989). Others as oleander (*Nerium oleander*), besides of causing poisoning of livestock (Pedroso et al. 2009) may represent a risky source of nectar or pollen to bees and provide more or less toxic honey to humans. Forest species may also become invasive and outcompete native plants or interfere with forestry operations, such as harvesting, or native forest regeneration. The kahili ginger (*Hedychium gardnerianum*) is a shrub of Asian origin which has been introduced in many countries for ornamental purposes. This plant has spread through human aid by horticultural industry and gardening activities but also by natural means through bird dispersion. In many countries, particularly in oceanic islands as the archipelagoes of Azores and Hawaii, the kahili ginger competes with native species and damages native forests preventing almost entirely their regeneration. Many exotic plants can be also toxic to animals as the black locust (*Robinia pseudoacacia*), an American tree introduced as ornamental in Europe where become invasive. In addition, invasive plants can attract unpleasant insects or pests or, when planted as hedge, can grow uncontrolled and cause disputes among neighbors.

Likewise invasive species of animals can provoke deep impacts on vegetation or cause soil erosion and are of especial concern to wellbeing because they are able to transmit parasites and fatal diseases (such as rabies) to humans and domestic animals. Many serious problems are the result of the escape or released of domestic animals introduced as livestock (Jaksic et al. 2002). Among these is the feral pig (*Sus scrofa*) knowing for causing important damage to domestic animals, crops and native biota in many parts of the world (Frederick 1998). Aquatic ecosystems are also very sensitive to the introduction of non-native species. Exotic aquatic species used as a recreational purpose for game or sport are considered responsible for reducing populations of many native species. In Europe invasive species of aquatic ecosystems are mainly crustacea (crabs, crayfishes) and mollusks (mussels, clam, oyster, snails).

6.2.2   Impacts of Invasive Species on Native Biodiversity and Ecosystems

The destructive impacts of biological invasions on native species, communities and ecosystems are less known to the general public. Nevertheless, in the last decades the effects of invasive species on biodiversity called the attention of scientists and their effects began being widely documented (Table 6.2). The impacts of invasive species can occur at population, at community and ecosystem levels but these last ones are still insufficiently understood (Kenis et al. 2009). At community level the effects of exotic species on the composition and functioning of invaded ecosystems can be diverse. Most information on ecosystem processes concern the impact caused by exotic plants but recent data suggest that even those caused by alien invertebrates can be dramatic (O'Dowd et al. 2003). The effects of invasive species to ecosystems may also be broad. Many non-native species can alter the physical habitat and modify the disturbance regimes (fire, erosion and flooding) or act as disturbance agents themselves. They may change the nutrient and water cycling, the energy flows through food webs and community structure (Charles and Dukes 2007). Even though these effects are remarkable, quantitative information for potential damages to ecosystems functioning and biodiversity are missing for most species.

At population level, the most well known impacts of alien species are a result of the effects on species interactions (competition, predation, herbivory, disease, pathogens), modification of dispersion patterns (e.g. dispersion of seeds by ants),

Table 6.2 Impacts of invasive species on natural ecosystems. (Adapted in part from Charles and Dukes 2007)

Impacts at community level	Impacts at population level
Affect species diversity (e.g. increase extinctions; decrease species diversity)	*Effects on species interactions*
	• Competition (e.g. allelopathy in plants)
	• Parasitism and pathogens
	• Mutalisms
	• Symbiosis
	• Predation
	• Herbivory
	• Fungivorous
	• Omnivorous
	• Pollinators
Alter community composition and interactions (e.g. add or modify functional groups; change vegetation structure)	*Hybridization with exotic species*
	Modification of dispersion patterns (e.g. of seeds; herbivores)
Change the natural cycles of energy, nutrients and water (e.g. decomposition rate; fixation of nitrogen)	
Modification of the disturbance regimes (fire, erosion and flooding)	
Modification of climate and atmosferic composition (e.g. alter emission of CO_2 or nitrogen)	
Changes physical habitat (e.g. secretion of salts; change pH)	*Habitat alteration*

habitat alteration, and the genetic effects (i.e. hybridization) (Christian 2001). One of the well-known impacts of exotic species on native organisms is competition either for habitat or resources. A remarkable example is the fast growing Australian acacia (*Acacia mearnsii*) one of the most harmful invasive species. This nitrogen-fixing tree causes an unforeseen chain of events on native habitats due to physical disturbances as soil erosion and water loss and to competition with indigenous vegetation which produce the substitution of grass communities and declining native biodiversity (Wilgen et al. 2001; Witkowski 1991).

Exotic herbivorous may also affect directly plant populations and cause profound impact and alteration of habitats. Widely known examples of herbivory effects are those caused by the goat (*Capra hircus*). Because this species ingests a great number of plant species is able to change plant communities and put at risk both plants (Mueller-Dombois and Spatz 1975) and animals due to reduction of shelter (Campbell and Donlan 2005). Another outstanding invasive species capable of causing severe soil erosion by overgrazing and burrowing is the rabbit (*Oryctolagus cuniculus*) (Courchamp et al. 2003). But goats, rabbits and other invasive species of vertebrates are also risky because they vectors of diseases and pathogens to native species (Pimentel 2007). On the other hand, exotic predators may kill and/or eat native species or may themselves sustain higher populations of native or non-native predators. There are examples of exotic predator species such as cats that attack native species (Frederick 1998). Domestic cats are responsible for the extermination of many native species of birds in many parts of the world, especially on remote habitats as islands in which native species evolved without any contact with predators (Kawakami and Fujita 2004).

Nevertheless, not all regions are equally vulnerable to the effects provoked by invasive species. Particularly isolated areas such as islands with endemic flora and fauna are far more sensitive to invasive species (Reaser et al. 2007). The number of invasions in islands is larger than in mainland. The reason for the high vulnerability of islands is not clear but several aspects such as the low diversity of native species, the missing of functional groups, a disharmonic community composition or a poor biotic resistance to competition and predators have been suggested (Simberloff 2000). The invasion on islands is particularly critical in Europe because although islands harbor a substantial part of the European biodiversity, they are also important stations for transoceanic shipping traffic, and as such, their vulnerability to the introduction of species from far away regions and heterogeneous sources, is usually much larger than mainland regions. This is the case of the Macaronesian archipelagos. The nearly 2.6 millions of inhabitants and 10 millions of tourists arriving each year demand a huge amount of natural resources and goods from other parts of the world. Due to tourism and welfare schemes, new exotic species associated with agriculture, gardening or recreational activities are continuously arriving leading to the proliferation of numerous exotic pests. In the Canary Islands alone about 43% of the 1,428 exotic terrestrial species were recorded in the last three decades (Martín et al. 2005). Unfortunately the present measures are insufficient to reduce this trend.

6.2.3   Which Exotic Species Have the Greatest Impact?

Not all species that arrive to a new area become established. It is estimated that only about 10% of the introduced species are able to build up new populations and of these 10% became invasive (Williamson 1996). The invasion process follows a pyramidal progression through which exotic species have to overcome several steps before becoming invasive. At first a species need to arrive to the new area and then has to be capable of establishing a population that sustains itself for many generations. Generally, the chance of establishment increases with the rate of arrival of an exotic species to a particular site. Once established, the species has to spread and persist before becoming a pest.

It is difficult to know which exotic species may become invasive and which not. In addition to the number of individuals released, another important factor is the capability of the habitat to support the development of the non-native species. Alien species more likely to become established are those capable of reproducing under various environmental conditions and utilizing different food sources (Long 1981). Within this group are many worldwide agricultural pest species as the tobacco whitefly (*Bemisia tabaci*), which causes destruction to many crops around the world. This insect, in addition of being able to adapt to a variety of host plants and unfavourable environmental conditions, has over 900 host plants mainly of agricultural interest and transmits 111 virus species (Jones 2003; Mound and Halsey 1978). However, some species are able do develop populations in new home areas even though they do not find suitable environmental conditions. This is the case of some species that can set up themselves in man-made habitats such as the West Indian dry wood termite, *Cryptotermes brevis*. This anthropophilic termite is believed to have been spreading since the beginning of Spanish Empire to the present time (Scheffrahn et al. 2009); in the new home areas it is found in houses but hardly ever occurs in trees or abandoned woods.

Other characters related to survival and adaptation is the level of genetic diversity of the introduced populations (Petit et al. 2004) or the dispersal capacity of individuals colonizing unoccupied habitats. The adaptability to different environmental conditions known as *ecological plasticity* is an important character to adjust rapidly and to survive in new environments. Numerous human commensals such as the house fly (*Musca domestica*), the common cockroach (*Periplaneta americana*) or social insects as ants, are highly adaptable to various environments like disturbed habitats and man's housing (Holway et al. 2002). Moreover, invasive species may gain competitive advantage over native species because they often leave behind in their native habitats their predators, parasites and diseases (Torchin et al. 2003). Successful invaders usually have more than one of these characteristics which facilitates them the successful colonization of new habitats. For example the worldwide distribution of the house mouse, *Mus musculus*, is facilitated among other traits by its commensal relationship with humans, a broad diet, being a prolific breeder and its ecological and behavioral plasticity. Since several traits may facilitate invasion, understanding the invasion process requires information on several characters as population biology, life history traits and genetic and evolutionary changes.

6.3   Vectors and Pathways Used by Invasive Species

Patterns of species introductions parallel the patterns of development of markets and trade routes of intercontinental commerce (Simberloff 1986). Therefore, affinity to humans and their goods is an important feature useful for screening potential invaders. The repeated routes by which invading species frequently arrive to an area are called *pathways*. The pathways associated with human activity are mainly related to commercial routes. The modes by which species are carried are called *vectors*, which may be intentional or unintentional (see Hulme et al. 2008). The intentional transport of exotic species may occur through commodities for food, agriculture, horticulture, biocontrol, medicine, game and pets. The unintentional transport of alien species may be in ship ballast (soil, water), transport containers and packaging or as contaminants of commercial products. Alien species arrive with the transportation of contaminated commodities, as stowaways in a transport vector and/or spread naturally from a nearby region where the species is alien.

The amount and the trend in the trade of particular commodities such fruits and alive plants may provide an alternative for estimating the potential for new exotic pests contaminants. However, not only the percentage of shipments affected is usually high but pests are also difficult to notice and even harder to predict. Yet, empty sea cargo containers, temporarily out of use, may have pests (Stanaway et al. 2001). This may provide us an approximate image of the huge problem that commercial activities represent. Therefore, species that are more likely to be transported as a result of this increased mobility should be considered high-hazard species.

Alive plant material arriving in commercial cargo is one of the most common vectors of exotic species. In addition to plant material either to be planted (nursery stocks, grafts, bulbs, tubers and seedlings) or to be used as ornamentals (cut flowers), there is a large flux of imported fruits and vegetables necessary for the needs of inhabitants and visitors. In ornamental plants the probability of invasion is related with horticultural trade (market and frequency of transport). In Britain demand for exotic plants considered 'good garden subjects' is known to be an important factor in the invasion success of introduced species (Dehnen-Schmutz et al. 2007). Many of these imported plants, in addition to becoming invasive, are able to introduce themselves an unnoticed number of associated pests. A study performed in Europe show that the main commodities on which pests arrived were cut flowers (22.3%), plants for planting and potted plants (19.1%) and vegetables (18.7%) especially bonsais (8.6%), wood/bark (3.7%) and wood derivates (2.3%) (Roques and Auger-Rozenberg 2006).

Stored products as cereal grains and wood products are also important sources of pest species particularly of invertebrates. As a consequence of the international trade of grains, pest fauna of stored products introduced with goods did also undergo a worldwide homogenization. Due to this, 26 pest species of stored products were excluded from the checklist of plant quarantine in 1998 (Kiritani and Yamamura 2003). Wood is also a common source of inadvertent introduction of pests especially in economically developed countries which are the largest importers of wood products (Laarman and Sedjo 1992; Tkacz 2002). Forest pests can be transported between countries and regions either through propagative materials or by wood products as logs or solid wood packaging (McCullough et al. 2006). An analysis of the regulated pests detected and published by EPPO for 29 European countries for the period between 1995 and 2004, showed that the interception of alien forest insects was much more important on wood packaging material (37.6%), and especially on casewood (34.0%), than on sawn timber (15.8%) and logs (8.1%) (Roques and Auger-Rozenberg 2006).

Domestic animals introduced as livestock often cause serious problems either because they escape or are released. Among the one hundred most invasive species are several domestic animals, such as the goat (*Capra hircus*), the cat (*Felis catus*), the rabbit (*Oryctolagus cuniculus*) and the feral pig (*Sus scrofa*) (GISD 2005). But also species introduced for fish farming can become invasive. The Chinese mitten crab (*Eriocheir sinensis*) initially imported legally in Europe and North America for human consumption and for aquariums cause presently damages of millions of euros to aquaculture, fisheries (Rudnick and Resh 2002) as well as to native species due to physical disturbance and ecosystem change. Animals imported as pets or as live bait and food are also major sources of invasive species to aquatic ecosystems (Courtenay 1999). The invasive alga, *Caulerpa taxifolia*, widely used as decorative plant in aquaria, is a major problem for marine life in the Mediterranean Sea (Meinesz and Hesse 1991). In addition, Alien organisms often bring to their new home areas parasites and devastating diseases and pathogens that spread around the world. Good examples of these are the exotic Newcastle disease, a contagious and lethal viral disease that affects all species of birds, or plant disease epidemics caused by fungal pathogens as potato blight, chestnut blight or Dutch elm disease (Brasier and Buck 2001).

In some cases intentional introduction of new species had the purpose of solving environmental problems but create additional ones. Biological control agents purposely released as natural enemies to manage the population density of invasive species can potentially cause collateral damage to indigenous species (Simberloff and Stiling 1996). Introduction of pollinators for increasing crop productions such as honeybees (*Apis mellifera*) and bumblebees (*Bombus terrestris*) may also represent a potential threat to native plants and pollinators (Velthuis and van Doorn 2006).

In addition to commercial routes, invasive species may arrive to a new region by non commercial activities as scientific or military. Military armed activities related to transport of people and equipment to and from distant locations as wars, peace-keeping, supplies in disaster/humanitarian actions and training exercises are well known for the spread of exotic species around the world (Westbrook and Ramos 2005; Vertegaal 1989). The brown tree snake (*Boiga irregularis*) is believed to have arrived in the Guam Island (western Pacific Ocean) as stowaway on a USA military transport by the end of World War II. Later on this small island this snake nearly eliminated the native forest birds (Fritts and Rodda 1998). Care should also be taken with aircrafts and ships. Insect vectors of human diseases particularly those of dengue, malaria or yellow fever, are introduced accidentally in this manner (Lovell and Drake 2009; Tatem et al. 2006). Several species of mosquitoes as *Aedes aegypti*, *A. albopictus* and *Culex pipiens* have been introduced in many areas by ship in car tires (Lounibos 2002).

6.4 Prevention of Invasive Species. A Challenge to Educators

A decrease in the entry of invasive species requires an increase in inspection and quarantine control together with cooperation between inspection agents and research institutions. The decrease of alien invasive species would be possible through a reduction of trade and travel; however, presently, this is not acceptable under our global economy. There is a clear conflict of interests between control of invasive species and global trade. Governments are responsible for providing regulations of public interest, but the present prevailing economic view of free trade favored by the policies of the World Trade Organization, requires the removal of restrictions to trade (inspection and quarantine), necessary to prevent the introduction of invasive species, undermining this way the capacity of governments to apply measures effectively (Low 2001). Therefore, economic, social and environmental costs caused by invasive species, considered by economists as "externalities", are suffered by others, usually farmers or general public different from those promoting introductions (McNeely et al. 2001).

Once an alien species is established it is difficult to eradicate, therefore, preventive measures are the most cost-effective means of control (IUCN 1999). Presently, international regulations on quarantine procedure deny the entry only when a species is proven to be noxious elsewhere. However, the most reliable attitude to avoid further biological invasions would be to consider the introduction of species as a threat until proven otherwise. Besides, it would be essential to implement appropriate legislation particularly at global scale following the principles of the Convention on Biological Diversity (McNeeley et al. 2001). Appropriate regulations on importation based on risk analysis and attributing management responsibilities to the commercial agents involved in the introduction of exotic species would be an important step.

Nevertheless, the effects of invasive species are so broad that cannot be of concern only to governments, scientists, or conservationists. Often people's perception of the status of introduced species is different from that of conservationists. Therefore, education of citizens particularly from early ages is a key process for increasing public awareness of invasive alien species. Many invasive species are the result of deliberate or involuntary introductions by citizens (Williamson 1996). The two main inconvenients that make communication difficult and limit public consciousness on invasive species are related to an incomplete understanding of people perceptions (Binggeli 2001) but also to the complexity of the topic itself. Biological invasions are a comprehensive issue which requires the understanding of fact and the scientific concepts involved. Lack of public awareness on invasive species may be related to the lack of basic knowledge such as how to differentiate between native or alien species (Colton and Alpert 1998).

But public perceptions regarding a species as valuable or undesirable may change also among societies and over the course of time depending on the individual or group of interests (Colton and Alpert 1998). A broad spectrum of perceptions and views related to invasive species are affected by traditions, beliefs and political views. The perceptions of a particular organism as a pest or weed depend very much on the interest of the observer and on the knowledge of their local environment (Edwards 1998; Wit et al. 2001). For example, invasive species of vertebrates raise usually confrontations among advocates of animal rights and other sectors of the society as hunters. Indeed, it is common that agriculturists, foresters, horticulturists, conservationists or scientists have different opinions on whether an introduced species is harmful or not. A combination of these different views can result sometimes in highly controversial and often irreconcilable behaviors and attitudes (Binggeli 1994). Because often debate is not focused on facts but on believes, discussions are sometimes almost impossible and may interfere with the actions to prevent and control invasive species (Binggeli 1994).

A critical understanding of scientific evidences by citizens is crucial to evaluate in each situation contrasting opinions of the different sectors and/individuals involved in order to take appropriate actions. Insufficient knowledge on the science of environmental issues is often responsible for persistent misconceptions (Gambro and Switzky 1999). Thus, schools need to include science-based curricula and address environmental issues (Osborne and Dillon 2008) and these should be reinforced in areas of special vulnerability to invasions, like islands. Understanding the effects, the routes of transport and control of invasive species is a cue to help reducing the magnitude of the problem in the future. Besides it would give support to the application of unpopular but important measures as inspections of hand luggage (Liebhold et al. 2006; Gratz et al. 2000) and quarantine inspections at ports-of-entry and country borders. Among some of the most common believes are that invasions happen all the time or, that humans have move at a planetary scale many species for thousands of years or, that too many people and commodities are moving around the world that little can be done. Therefore, engaging students from early ages is critical to change this hopeless picture regarding invasive species prevention.

In formal contexts at school, students can be involved in practical activities useful to understand scientific concepts but also may participate in informal out school activities as habitat restoration or removing of invasive species in nearby areas, which can assist to promote a change in attitude. The use of daily life examples that involve a risk of introduction of invasive species is helpful for discussion/debates or hands-on-activities (Table 6.3). Appropriate examples of unintended (postal services or tourism) and intended introductions (trade of pets, flowers or alive plants) can be chosen according to the community in which the school is involved. Many exotic specimens dead or alive like seeds and animals are available via internet base companies and often enter unnoticed through mail that seldom is examined (Kay and Hoyle 2001).

Dangers associated to the introduction of exotic species can be addressed using well known icons to students as Christmas trees. Lessons and other activities on invasive species can be suitable for periods of the year where travelling is more likely and the possibilities to buy and bring foreign items home increase. Illegal commerce and traffic of wild animals are also a major source of introduce invasive species and this is an important issue to be address at school since children are often targeted as consumers of exotic pet animals. The invasion success of many pet animals like parrots (Cassey et al. 2004) and aquarium fish (Duggan et al. 2006) is related to the availability and frequency of species in trade. Many of these species do often escape or are abandoned by their owners being a serious danger to humans and wildlife. Science lessons addressing animal welfare focusing on care and maintenance needs and on the stress and anxiety that animals suffer when kept under inappropriate living conditions, will also assist to educate more sympathetic citizens. Although the challenges faced by science education are to educate future citizens capable to undertake responsible participation in science-based issues, science is often perceived by students as boring, outdated or irrelevant (Sjøberg 2003). Global environmental issues such as invasive species demand, as never did before, an urgent collaborative work among teachers and scientists to construct motivated and meaningful curricula.

Table 6.3 Some guidelines on actions that citizens can undertake to prevent the introduction and spreading of exotic species

Plant material

- Use for landscape and gardening plants native to your area
- Select alive material (fruits, vegetables, plants, flowers, christmas trees, etc) prefereable produced in areas close where you live
- Choose wood material (firewood, wood for fences, etc) grown preferable in close regions to your home area and, if buying from far regions, ask for a importation certificate
- Avoid planting invasive species and, if you have invasive species, prevent them from spreading beyond your garden
- Try not to buy and plant mixtures of seeds
- Do not throw undesirable plants or garden cuttings in natural areas

Pet animals

- When buying pets or domestic animals (birds, fishes, dogs, cats, insects, etc) deparasitize them to reduce the risk of introduction and spreading of new parasites
- Look after yourself and your pets to avoid new diseases and vectors
- Try not to have exotic animals as pets and, if you buy them, select those that were breed in captivity in your area
- If you buy an exotic pet do it from reliable dealers whose animals are legally imported, and ask always for the selling permission
- Avoid to take pet animals to protected areas (e.g. natural parks)
- Never abandon pet animals in nature or aquarium fish into a natural body of water
- When cleaning pet animals and their associated materials (food dishes, etc) be sure that you do not throw with it alive material or parasites in nature
- Do not discard aquatic plants or aquarium water into local waters

Tourism and travelling

- Never smuggle or carry alive material (plants, fruits, seeds, soil or animals) between different regions
- Avoid to transport fruits or vegetables while you are travelling between different countries
- Inspect and clean your hand luggage, clothes (boots, etc) and other items (cars, bikes, etc) from caked-on soil and seeds before and after traveling
- Do not transport items such as food, wood, soil, grass or gravel from one part to another
- Follow quarantine regulations to prevent the spread of invasive species, parasites and diseases
- When visiting natural areas do not take alive materials (plants, fruits or animals) with you

Boating and fishing

- Do not transport water, animals, or plants from one water source to another
- Remove all suspicious aquatic material (plants and animals) from your goods and equipment before leaving to a new area
- Wash always boats and other equipment before leaving to a new area

Education

- Read information about invasive species and make an effort to understand the scientific concepts involved and their impacts
- Learn to recognize invasive and native species present in your area
- Participate in eradication campaigns and monitoring activities of invasive species
- Share your knowledge on invasive species with others

Acknowledgment I thank Ana Cunha for her comments on this manuscript.

References

Binggeli P (1994) The misuse of terminology and anthropomorphic concepts in the description of introduced species. Bull Br Ecol Soc 25:10–13

Binggeli P (2001) Human dimensions of invasive woody plants. In: McNeely JA (ed) The great reshuffling: human dimensions of invasive alien species. IUCN, Gland

Brasier CM, Buck KW (2001) Rapid evolutionary changes in a globally invading fungal pathogen (Dutch elm disease). Biol Invas 3:223–233

Bright C (1999) Invasive species: pathogens of globalization. Foreign Policy Fall 1999:50–64

Campbell KJ, Donlan CJ (2005) A review of feral goat eradication on islands. Conserv Biol 19:1362–1374

Cassey P, Blackburn TM, Russell GJ, Jones KE, Lockwood JL (2004) Influences on the transport and establishment of exotic bird species: an analysis of the parrots (Psittaciformes) of the world. Glob Change Biol 10:417–426

Charles H, Dukes JS (2007) Impacts of invasive species on ecosystem services. In: Nentwig W (ed) Biological invasions. Springer, Berlin

Christian CE (2001) Consequences of a biological invasion reveal the importance of mutualism for plant communities. Nature 413:635–638

Colton TF, Alpert P (1998) Lack of public awareness of biological invasions by plants. Nat Areas J 18:262–266

Convention on Biological Diversity (CBD) (2000) Global strategy on invasive alien species. Convention on Biological Diversity, UNEP/CBD/SBSTTA/6/INF/9

Courchamp F, Chapuis J-L, Pascal M (2003) Mammal invaders on islands: impact, control and control impact. Biol Rev 78:347–383

Courtenay W. R. Jr (1999) Aquariums and water gardens as vectors of introduction. In: Claudi R, Leach J (eds) Nonindigenous freshwater organisms: vectors, biology and impacts. Lewis, Boca Raton

Crosby AW (1986) Ecological imperialism. Cambridge University Press, New York

Dehnen-Schmutz K, Touza J, Perrings C et al (2007) A century of the ornamental plant trade and its impact on invasion success. Divers Distrib 13:527–534

Di Castri F, Hansen AJ, Debussche M (1990) Biological invasions in Europe and the Mediterranean Basin. Kluwer Academic, Dordrecht

Ding J, Mack RN, Lu P, Ren M, Huang H (2008) China's Booming economy is sparking and accelerating biological invasions. Bioscience 58:317–324

Duggan IC, Rixon CAM, MacIsaac HJ (2006) Popularity and propagule pressure: determinants of introduction and establishment of aquarium fish. Biol Invas 8:377–382

Edwards K (1998) A critique of the general approach to invasive plant species. In: Starfinger U, Edwards K, Kowarik I, Williamson M (eds) Plant invasions: ecological mechanisms and human responses. Backhuys, Leiden

European Commission (2006) Communication of the commission 'Halting the loss of Biodiversity in 2010'—and beyond. Sustaining ecosystem services for human well-being

Frederick J (1998) Overview of wild pig damage in California. Vertebr Pest Conf 18:82–86

Fritts TH, Rodda GH (1998) The role of introduced species in the degradation of island ecosystems: a case history of Guam. Annu Rev Ecol Syst 29:113–140

Gambro JS, Switzky HN (1999) Variables associated with American high schools students' knowledge of environmental issues related to energy and pollution. J Environ Educ 30:15–22

Global Invasive Species Database (GISD) (2005) Global invasive species database. http://www.issg.org/database. Accessed 03 Nov 2010

Gratz NG, Steffan R, Cocksedge W (2000) Why aircraft disinfection? Bull World Health Organ 78:995–1004

Holway DA, Lach L, Suare AV et al (2002) The causes and consequences of ant invasions. Annu Rev Ecol Syst 33:181–233

Hulme PE (2007) Biological invasions in Europe: drivers, pressures, states, impacts and responses. In: Hester R, Harrison RM (eds) Biodiversity under threat issues in environmental science and technology. Royal Society of Chemistry, Cambridge

Hulme PE, Bacher S, Kenis M et al (2008) Grasping at the routes of biological invasions: a framework for integrating pathways into policy. J Appl Ecol 45:403–414

IUCN (World Conservation Union) (2000) IUCN Guidelines for the prevention of biodiversity loss caused by alien invasive species. IUCN, Gland, p 24

Jaksic FM, Iriarte JA, Jiménez JE et al (2002) Invaders without frontiers: cross-border invasions of exotic mammals. Biol Invas 4:157–173

Jones D (2003) Plant viruses transmitted by whiteflies. Eur J Plant Pathol 109:197–221

Kawakami K, Fujita M (2004) Feral cat predation on seabirds on Hahajima, the Bonin Islands, Southern Japan. Ornithol Sci 3:155–158

Kay SH, Hoyle ST (2001) Mail order, the internet, and invasive aquatic weeds. J Aquat Plant Manag 39:88–91

Kenis M, Auger-Rozenberg MA, Roques A et al (2009) Ecological effects of invasive alien insects. Biol Invas 11:21–45

Kiritani K, Yamamura K (2003) Exotic insects and their pathways for invasion. In: Ruiz G, Carlton JT (eds) Invasive species: vectors and management strategies. Island Press, Washington

Laarman JG, Sedjo RA (1992) Global forests. McGraw-Hill, New York

Liebhold AM, Work TT, McCullough DG et al (2006) Airline baggage as a pathway for alien insect species invading the United States. Am Entomol 52:48–54

Long JL (1981) Introduced birds of the world. Reed, Wellington

Loope LL, Howarth FG (2002) Globalization and pest invasion: where will be in five years? In: Proceedings of the 1st International Symposium on Biological Control of Arthropods, Honolulu, Hawaii, 14–18 Jan 2002, pp 34–39

Lounibos LP (2002) Invasions by insect vectors of human disease. Annu Rev Entomol 47:233–266

Lovell SJ, Drake LA (2009) Tiny stowaways: analyzing the economic benefits of a U.S. environmental protection agency permit regulating ballas. Water discharges. Environ Manag 43:546–555

Low T (2001) From ecology to politics: the human side of alien invasions. In: McNeely JA (ed) The great reshuffling: human dimensions of invasive alien species. IUCN, Gland

Martín JL, Marrero MC, Zurita N et al (2005) Biodiversidad en gráficas. Especies silvestres de las Islas Canarias. Consejería de Medio Ambiente y Ordenación Territorial. Gobierno de Canarias, Santa Cruz de Tenerife

McCullough DG, Work TT, Cavey JF et al (2006) Interceptions of non-indigenous plant pests at US ports of entry and border crossings over a 17-year period. Biol Invas 8:611–630

McLennan MW, Amos ML (1989) Treatment of lantana poisoning in cattle. Aust Vet J 66:93–94

McNeely JA, Mooney HA, Neville LE et al (2001) A global strategy on invasive alien species. IUCN, Gland

Meinesz A, Hesse B (1991) Introduction of the tropical alga *Caulerpa taxifolia* and its invasion of the northwestern Mediterranean. Oceanol Acta 14:415–426

Mooney HA, Drake JA (1989) Biological invasions: a SCOPE program overview. In: Drake JA, Mooney HA, di Castri F, Groves RH, Kruger FJ, Rejmánek M, Williamson M (eds) Biological invasions: a global perspective. Wiley, New York

Mound LA, Halsey SH (1978) Whiteflies of the world, a systematic catalogue of the Aleyrodidae (Homoptera) with host plant and natural enemy data. British Museum (Natural History), London

Mueller-Dombois D, Spatz G (1975) The influence of feral goats on the lowland vegetation in Hawaii Volcanoes National Park. Phytocoenologia 3:1–29

Norton DA (2009) Species invasions and the limits to restoration: learning from the New Zealand experience. Science 325:569–571

O'Dowd DJ, Green PT, Lake PS (2003) Invasional meltdown on an oceanic island. Ecol Lett 6:812–817

Osborne J, Dillon J (2008) Science education in Europe: critical reflections. A report to the nuffield foundation. http://www.nuffieldfoundation.org/fileLibrary/pdf/Sci_Ed_in_Europe_Report_Final.pdf

Pedroso PMO, Bandarra PM, Bezerra Júnior PS et al (2009) Natural and experimental poisoning by *Nerium oleander* (Apocynaceae) in cattle in Rio Grande do Sul. Pesq Vet Bras 29:404–408

Petit RJ, Bialozyt R, Garnier-Géré P et al (2004) Ecology and genetics of tree invasions: from recent introductions to quaternary migrations. For Ecol Manag 197:117–137

Pimentel D (2007) Environmental and economic costs of vertebrate species invasions into the United States. In: Witmer GW, Pitt WC, Fagerstone KA (eds) Managing vertebrate invasive species: Proceedings of an International Symposium USDA/APHIS/WS, National Wildlife Research Center, Fort Collins 2007. http://www.aphis.usda.gov/wildlife_damage/nwrc/symposia/invasive_symposium/content/Pimentel2_8_MVIS.pdf

Pimentel D, Zuniga R, Morrison D (2005) Update on the environmental and economic costs associated with alien-invasive species in the United States. Ecol Econ 52:273–288

Reaser JK, Meyerson LA, Cronk Q et al (2007) Ecological and socioeconomic impacts of invasive alien species in island ecosystems. Environ Conserv 34:98–111

Rees M, Hill RL (2001) Large-scale disturbances, biological control and the dynamics of gorse populations. J Appl Ecol 38:364–377

Richardson DM, Pysek P, Rejmánek M et al (2000) Naturalization and invasion of alien plants: concepts and definitions. Divers Distrib 6:93–107

Roques A, Auger-Rozenberg MA (2006) Tentative analysis of the interceptions of nonindigenous organisms in Europe during 1995–2004. EPPO Bull 36:490–496

Rudnick DA, Resh VH (2002) A survey to examine the effects of the Chinese mitten crab on commercial fisheries in northern California. Interag Ecol Program Newsl 15:19–21

Scheffrahn RH, Křeček J, Ripa R et al (2009) Endemic origin and vast anthropogenic dispersal of the West Indian drywood termite. Biol Invasions 11:787–799

Simberloff D (1986) Introduced insects: a biogeographic and systematic perspective. In: Mooney HA, Drake JA (eds) Ecology of biological invasions of North America and Hawaii. Springer, New York

Simberloff D (2000) Extinction-proneness of island species—Causes and management implications. Raffles Bull Zool 48:1–9

Simberloff D, Stiling P (1996) How risky is biological control? Ecology 77:1965–1974

Sjøberg S (2003) Science and technology education current challenges and possible solutions. In: Edgar J (ed) Innovations in science and technology education, vol VIII. UNESCO, Paris

Stanaway MA, Zalucki MP, Gillespie PS et al (2001) Pest risk assessment of insects in sea cargo containers. Aust J Entomol 40:180–192

Tatem AJ, Hay SI, Rogers DJ (2006) Global traffic and disease vector dispersal. Proc Natl Acad Sci U S A 103:6242–6247

Tkacz BM (2002) Pest risks associated with importing wood to the United States. Can J Plant Pathol 24:111–116

Torchin ME, Lafferty KD, Dobson AP et al (2003) Introduced species and their missing parasites. Nature 421:628–630

Valéry L, Fritz H, Lefeuvre J-C et al (2008) In search of a real definition of the biological invasion phenomenon itself. Biol Invasions 10:1345–1351

Velthuis HHW, van Doorn A (2006) Apidologie. A century of advances in bumble bee domestication and the economic and environmental aspects of its commercialization for pollination. Apidologie 37:421–451

Vertegaal PJM (1989) Environmental impact of Dutch military activities. Environ Conserv 16:54–64

Webb DA (1985) What are the criteria for presuming native status? Watsonia 15:231–236

Westbrook C, Ramos K (2005) Under siege: invasive species on military bases. NationalWildlife Federation. http://www.necis.net/necis/files/nwfunder_seige_1005.pdf. Accessed 15 Oct 2010

Williamson M (1996) Biological invasions. Chapman and Hall, London

Wilgen van BW, Richardson DM, Le Maitre DC et al (2001) The economic consequences of alien plant Invasions: examples of impacts and approaches to Sustainable management in South Africa. Environ Dev Sustain 3:145–168

Wit MP de, Crookes DJ, Wilgen van BW (2001) Conflicts of interest in environmental management: estimating. Biol Invas 3:167–178

Witkowski ETF (1991) Effects of invasive alien Acacias on nutrient cycling in the coastal lowlands of the Cape Fynbos. J Appl Ecol 28:1–15

Work TT, McCullough DG, Cavey JF et al (2005) Arrival rate of nonindigenous insect species into the United States through foreign trade. Biol Invas 2:323–332

Chapter 7
Psychology for the Sake of the Environment

Joan Freeman

Humankind cannot bear very much reality.
(T. S. Elliott)

Abstract As people begin to take responsibility for the welfare of the earth and no longer see themselves as passive spectators of uncontrollable climatic forces, psychology can offer support and direction. Psychological techniques can help to change both attitudes and behaviour for the benefit of the environment. The greatest influences for positive change usually come from the social and educational psychology of the home, school and workplace.

Children can be educated to see beyond their domestic world, and draw on their learning so as to use it for the community's mutual benefit. Consideration of difficult concepts such as time, space, biodiversity, the troposphere and technology calls for an intelligent and educated population which can read and think scientifically to distinguish genuine evidence from false trails and conspiracy theories. This chapter looks at psychological research evidence and draws practical conclusions from it on changing minds and behaviour for the sake of the environment.

To change habits, we need to know, for example, why people say out loud that they want to stop negative climate change and yet behave recklessly in climate terms in their everyday lives. There are conflicts of interests to be addressed here.

7.1 The Psychological Challenge

Scientific and individual beliefs on global warming, pollution and the loss of natural resources are subjects of heated contention around the world. Every day, the media report catastrophic global climate changes of many kinds to which people react in a variety of ways. The debates concern events beyond the immediate environment, taking in the whole planet, including the troposphere—the atmosphere that supports all life.

J. Freeman (✉)
Middlesex University, 21 Montagu Square, W1H 2LF London, UK
e-mail: joan@joanfreeman.com
www.joanfreeman.com

A. Mendonca et al. (eds.), *Natural Resources, Sustainability and Humanity,*
DOI 10.1007/978-94-007-1321-5_7, © Springer Science+Business Media Dordrecht 2012

In the more developed parts of the world, concepts such as sustainability and biodiversity along with aiming to be 'green' have become part of everyday conversation, and almost a social norm. But this does not include everyone. In Britain, for example, those most likely to make positive ecological changes to their life-styles are aged over 65, live in rural areas or are of a higher social class (Survey for the Environment Food and Rural Affairs Department, Nov 2007 Tinyurl.com/ypzenk). These data show the need to involve younger people, those in cities and the less well informed.

Scientists generally work at two ends of a spectrum. At one end there are 'hard' scientists (physicists, chemists etc) and at the other end the 'soft' scientists (anthropologists and psychologists etc). Yet workers at the two ends of the spectrum do overlap to work together. The 'hard' scientists investigating climate change in the physical world constantly present overwhelming evidence of the steady destruction of the earth's bounties. They measure what it is happening and offer reasons why. 'Soft' scientists of the psychological world show that if there is to be a positive change people can no longer see themselves as passive spectators of uncontrollable forces, but argue that they must take responsibility for the welfare of the planet. This chapter aims to show how this might be done.

As Ivan Pavlov showed in his classical conditioning experiments in Leningrad in the 1920s, the best way of changing behaviour is to follow any stimulus as quickly as possible with a response, whether reward or punishment. But climate change is more complicated. The problems are in deeply entrenched habits of thought, notably in the concepts of time and space, which are extremely difficult to reach and change. Of itself, longer-term thinking and planning is much harder to do than producing a short term reaction. In climate terms, greenhouse gases have such a long life that to control them means planning for hundreds not tens of years. What is happening to the climate now has come from actions taken by past generations, and the long time-lag ahead means that any benefits of people's current efforts will not be seen until long after the present population is dead.

To start changing perceptions and behaviour, it is necessary to understand the old familiar ways, which are as basic as seeing and hearing. For example, you might be very familiar with a field nearby because you have grown up close to its stillness and beauty with only the sound of the birds, until one day workmen arrive and churn it up, bringing great anxiety, leaving it filled with enormous, white, turning windmills, cutting the air and filling the sky with a terrible noise. What has been a familiar sensual pleasure, knowing and loving that land, has been drastically destroyed and for the rest of your life. Nobody asked you—a faceless bureaucracy did it. Your head knows that the windmills are there to grab energy from the wind, but your heart is angry. Emotionally it is almost impossible to coordinate your love of the natural world with such machinery designed to preserve it.

One major problem in the everyday world is the difficulty most people have in understanding what the scientists are telling them about climate change. They are too often, it seems, talking from another planet! If people are to adopt a different way of living, it is important that information is presented in an understandable way, and that instructions are not only clear and easy to follow but also visually easy to read. Without that clarity and the possibility of a real understanding of what is needed, it may all seem too difficult and pointless.

David MacKay, a physicist at Cambridge University, has attempted to explain things in normal language (MacKay 2009). He points out, for example, how well-meaning media which urge the public to change their behaviour can even be misleading.

> … the idea that one of the top ten things you should do to make a difference to your energy consumption is to switch off the phone-charger when you are not using it. The truth is that leaving the phone charger switched on uses about 0.01 kWh per day. This means that switching the phone charger off for a whole day saves the same energy as is used in driving an average car for one second. Switching off phone chargers is like bailing the Titanic with a teaspoon.
>
> … all hydrogen-powered transport prototypes increase energy consumption compared to ordinary fossil-cars; whereas electric vehicles are significantly more energy efficient than fossil-cars. So hydrogen vehicles make our energy problem worse, and electric vehicles make it better.

Although, as MacKay says, the effect of a single individual's actions in saving energy will not make any difference to the whole world, the effect of many millions saving energy could be significantly effective. For example, switching from gas-guzzling to fuel-efficient cars in the US alone would nearly offset the emissions generated in providing electricity to 1.6 billion people.

The Swiss psychoanalyst, Jung (1964), described how people often erect psychological barriers to protect themselves from "the shock of facing something new", due to a "deep and suspicious fear of novelty". To preserve the warm feelings of familiarity, most of us naturally aim at finding cracks in evidence that suggests we should do something we don't want to do.

Psychologists know that simply telling people to do the right thing is unlikely to change their behaviour. It is much more effective to change administrative policy. An example is the current attempts in the over-eating Western world to change school-children's diets to diminish the growing epidemic of obesity and ill-health. Simply telling children not eat junk food has been found to be generally a waste of time, but if the school administration bans it from schools and replaces it with healthy food, the children's quality of eating significantly improves.

The psychological key is in changing attitudes which in turn changes behaviour. It is to do with moving away from relying on personal experiences toward the acceptance of scientific evidence. The aim is to change the action of each individual by their acceptance of the evidence. Acceptance is typically a three stage process. Denial and apathy come before acceptance.

7.1.1   Stages of Acceptance

Denial of the Problem Initial denial is driven by a mental phenomenon which psychologists call 'normalcy bias'. Denial comes from both lack of knowledge and more importantly experience. For example, people who have never experienced a catastrophe have difficulty recognising the signs that something awful is about to take place. Survivors of catastrophes very often tell how the non-survivors could not believe what was happening, were not flexible enough to adapt to the new situation, and so did not act appropriately to save themselves.

Ben Goldacre, the popular British science writer, refers to "zombie arguments" (Goldacre 2008). These are arguments which should have been killed off by the evidence against them, but they survive and are raised again. Zombie arguments are resistant to death because they neither live nor die by the normal standards of human rationality. Typical undead arguments in climate change are that "CO_2 is not an important greenhouse gas", "Global warming is down to the sun", and "Climate has always changed over the centuries, so there is nothing new".

Apathy in the Face of Challenge Psychologically, when facing challenge, it is much more effective to be part of the solution rather than a bystander. Accepting the situation and going along with it does not change anything. There is plenty of evidence that the most stressful part of a challenge is anticipation, uncertainty and the fear of what is likely to happen. The best way to start moving out of a fearful situation, such as the threat of climate change, is to know as much as possible about the threat, and accept it as reality. For both children and adults, it is possible to release anxiety in the face of change by using strategies of analogy—making the unfamiliar familiar. In children, this is called play; in adults, it is called rehearsal.

Taking Action The most difficult part of being in a threatening situation, such as the drying up of natural resources, means making an effort to relieve the anxiety and possibly stop the threat. The reward is that taking action resolves tension and brings relief—a very positive feeling. Taking action also means that it's easier to stop thinking about your internal experience of fear and instead focus usefully on external things, such as improving your situation. The psychological challenge is to move people from denial, and apathy to being in a state of willingness to take action and then to act.

7.2 Psychology for Climate Change

Environmental psychology has moved fast from a focus on the built environment at the end of the twentieth century to that of the natural environment (Cassidy 1997; Nickerson 2002). Most particularly, it is attempting to understand the way we live socially, which affects who we believe we are and what we are entitled to. The work-place adds to the influences of other human factors involved in the development and use of technology for energy-efficiency and recycling. Then there is consumerism, risk assessment and cost-benefit analysis. It is impossible to live an entirely green life. This means that people who are concerned for the environment face never-ending, unsolvable dilemmas and restrictions on their basic wish to live a happy stress-free life.

Some pertinent questions for psychology

- How can we hold two or more inconsistent ideas in our heads at the same time?
- Why do people say one thing and do another?
- Why do people behave inconsistently from one situation to another?
- How do people translate their feelings and beliefs into actions?

The constant problem for all psychology is that because it is concerned with human beings it is far from being an exact science. Psychological problems are rarely clear-cut with only one right solution: sometimes several seem equally right. Solutions may even demand an intuitive approach with a mixture of information and feelings. Choices in everyday dilemmas, like who you marry or where you choose to live, will change your life and happiness and are often based largely on intuition. In fact, just identifying the cause of an everyday problem, let alone the solution, is difficult because it is part of the way you live. Everyday dilemmas are also persistent; one decision sometimes only seems to pave the way for a new one. Furthermore, solving a life problem is one thing, but convincing other people of the rightness of your solution is another.

7.2.1   Cognitive Dissonance and Reward

The idea of holding clashing ideas in one's head at the same time was first presented by Leon Festinger in his theory, "Cognitive Dissonance" (Festinger 1951). The dissonance part, he said, "is experienced as uncomfortable tension", in spite of which we all hang on to that tension by acting against our beliefs at times. Perhaps as you put a cigarette to your lips, you have decided to live with the idea that smoking is bad for you. Or maybe, as you reach for one more cream cake, you push aside thoughts of your high cholesterol level. Everyone understands the meaning of Cognitive Dissonance because we all do it (Cooper 2007).

The way out of that tension is in making a decision to release it. You decide not to eat the cream cake. With that decision, the tension of Cognitive Dissonance, of knowing the facts about cholesterol yet anticipating the taste, is removed (if only for the time being). Your reward of emotional satisfaction supports your decision. In fact, some people may deliberately create that tension to get the emotional reward that follows. Playing with tension is a human trait and may, in fact, be at the roots of curiosity and the need for variety.

Another reward route to change is incentive-driven strategies. Countries such as Mexico and Brazil and trials by the Mayor of New York, Michael Bloomberg, are experimenting with them. These conditional transfers, as they are called, pay or reward individuals to change their ways. This might be rewarding children in cash for getting to school on time or parents for getting their children immunised.

But others find that the conflict of choosing is too painful. They do their best to avoid it, closing their minds to questioning and doubt. The simplest way of dealing with what we do not want to see is to deny it. Many do not see, for example, that current climate changes may be increasing to dangerous levels because of human behaviour. No, they argue, climate has always changed across the millennia. They deny responsibility, claiming that there is nothing we can do about it. Anna Freud called such attitudes "defence mechanisms" (Freud 1937). She wrote that defensive emotional strategies are created when people are confronted by an anxiety-provoking situation and unconsciously avoid dealing with it.

A psychological defence may be seen simply as *mañana*, putting things off, or refusing to face change by arguing that it has always been done that way. Then along come the scientists who show the need for a change in behaviour and challenge the way things are. A defence could be, for example, by the owners of water companies which have many big leaks in the system, but who fear the financial cost of stopping them. It is so much easier for the company to deny what is happening and its wasteful effects on natural resources. Psychological understanding aims to recognise the reasons for such defensive barriers to change and point to ways in which they may be overcome.

These deep barriers are in addition to those of apathy and inertia. On the whole, humans prefer do nothing. One way out of this is further administration, such as making consent the default-option. So, for example, instead of asking people to volunteer their organs for donation to others on death, some countries and several American states are making organ donation the default option. People have to make the effort to state specifically in writing that they do not want their organs used for others when they die. Without their effort to make that statement, their inertia can be used for the good of others.

7.3   Psychological Capital—Intuition and Culture

Feelings guide our actions, perhaps more than we would like to think, because none of us can be experts in every decision we make (Hogarth 2001). We make daily intuitive choices of what feels right in the situation, though we cannot explain why. How, for example, did you decide what to wear today? Which garment, and why? Why did you automatically greet one person but hesitate before speaking to another? The trouble is that intuition is unreliable: the fact that some decisions are right in one situation does not guarantee they will be right in another.

Cultural influences on individuals have deep historical origins along with mythology and religion. These effects can be seen, for example, in divisions of work in roles prescribed for social-classes or gender. Culture filters through generations when parents teach their children how to behave, but it also spreads horizontally, as when a dominant culture will affect others, such as the current world-wide American influence. Cultural influences also come from creative endeavour, for example the psychological ideas of Sigmund Freud or Pablo Picasso's concepts of art. With all these currents and cross-currents, the culture inherited by a particular generation is never the same as the one it passes on.

Cultural beliefs strongly influence a major environmental problem—over-population—too many mouths to feed. Maybe condoms are the greenest technology of all. If the population keeps increasing as it is today, we will need a second earth to sustain the coming generations. But this rarely features on the agenda of any agency aiming for climate control. Only in China, where the one-child policy may have led to 300–400 million fewer people being born, is population control seen as crucial to curbing emissions. China's population is expected to peak at around 1.4 billion in 2020, whilst that of India continues to grow swiftly. Mrs Gandhi's idea of transistors for every man who was sterilised was good in its time, but that project had problems of implementation and there has not been any follow up to it.

Birth rate, gender equality, education and poverty are inextricably linked. More than 200 million women worldwide have no access to contraception. It is widely accepted that women's education is the key to a lower birth rate, improved child health and a higher standard of family living; the more girls go to school and the more women who are employed, the fewer children there will be.

Our personal psychological capital emerges in our intuitions which, along with our personalities, work within our culture. The Russian, Lev Vigotsky, was the first to recognise this effect in his 'socio-historical' approach (Vigotsky 1978). He pointed out that while children are learning to speak, they are also taking in 'ready made' parcels of culture which affect all their communicating and thinking. The system works, he wrote, because adults in the culture have learned it and share the cultural assumptions. To change people's attitudes for the sake of the environment, psychology has to recognise and deal with this deep and powerful cultural influence—both within each individual and in the society.

Family and environmental cultural influences were clear in data from research I started in 1974 in Britain. I was investigating the experiences and outlooks of bright young people (N = 169, mean IQ 135, mean age 18) and their parents (Freeman 1991, 2010). I asked them about the prospect of a nuclear holocaust, how it might start and what might happen. The question produced impassioned responses which were analysed in terms of their measured IQs, education, upbringing and personalities. Strikingly, their IQ scores were associated with their attitudes at a very high level of significance (0.01). Those with the highest IQs were more inclined to believe in possible man-made destruction than the lower scorers who were more likely to believe that some outside entity, or god, would save them.

There were no differences in responses, though, in terms of age or gender, sensitivity to their fellow humans, or whether they had more troubled personal lives. Those who anticipated disaster were more likely to come from higher social-class, better educated families, right back through to grandparents, though there were no differences in their physical home and neighbourhood circumstances. The family differences were clearly not to do with money, but with behaviour and outlook. It is parents who teach their children that they are effective and competent in dealing with life.

The brightest and most highly educated young people were the liveliest thinkers and the ones most likely to take action. But they were also the pessimists who had a more heightened awareness and concern for the society they lived in. They were also twice as likely to be first-borns. The optimists protected themselves with psychological defence mechanisms, notably of two types—either the some higher authority would come to their aid and prevent destruction, or there really was no nuclear threat.

7.4   Smoking—An Example of Successful Changes of Mind and Behaviour

The rise of anti-smoking feeling and widespread action against smoking provides an excellent example of how attitudes and behaviour can be changed. The highly successful key has been in efforts to approach people's psychological capital. In the middle of the twentieth century, such a change in behaviour seemed an impossible goal. The tobacco industry had infinitely more resources than the tiny sums the health education campaigners could raise. Smoking advertisements were everywhere—in the media and on the streets—while their advertising jingles rattled on in the mind. Today, it is the same for ecology. For example, the estimated budget for Greenpeace is about $ 20 million a year, while that for advertisers of consumables worldwide is probably around $ 400 billion.

But in addition to lack of means, the health educator's major thrust for many years was simply to tell people how bad smoking was for them. Psychologically, like the notices in the doctor's waiting room, it had no recogniseable effect. The assumptions of the time about the nature of smoking: that it was normal, sophisticated—and a human right—seemed unchallengeable. When news of the ill-effects of tobacco began to be made public in the 1960s, there was a famous quote by an American tobacco executive. He said that "doubt is our product", meaning that they were no longer only selling tobacco, but also uncertainty, promoting the thought that maybe tobacco was not really poisonous, in spite of the scientific evidence. In the same way today, some still refer to climate change as though it were merely a possibility.

Although study after study published by scientists showed the benefits of cutting out smoking, there was no change of minds or fashion until there was real leadership in the form of government edicts. To start with, cigarette advertising was banned. Now, as countries rush to ban smoking in enclosed public places (and some in the open air) the positive effects can be seen in quality of life.

The increasing enactment of a smoking ban is possible because of the steep rise in public understanding of the effects of smoking. And smoking levels continue to go down (Office for National Statistics 2008). In 2008 only 22% of Britons aged over 16 smokes, down from 24% the year before and from 45% in 1974. Strangely, more girls are now smoking (10%) than boys (7%), for which there is no explanation. Looking back over half a century there has almost been a reversal of belief in the social value of smoking. What lessons from that successful campaign could be applied to the much less personal effects of world climate control?

7.4.1 Four Lessons from the Anti-Smoking Campaign for Climate Control

Lesson 1: Challenge It is possible to challenge deep assumptions—the psychological capital—of vast populations of all ages and from many cultures. Challenge to beliefs can open the possibility of change. As with smoking, concern about climate change raises three challenging questions which need resolving before many would be prepared to change their assumptions and habits.

1. What is true and what is not true?
2. What are the immediate benefits to the individual as well as to the wider world?
3. What can each individual do about it?

Evidence must be offered in a language which is easily understood and persuasive, and from a trustworthy source. Sometimes the information does not always clarify the issues and may lack conviction. Television and films are easily accessible, as is the internet, notably in the world-wide interactives such as Facebook, Myspace or personal blogs.

Psychology can help to get messages through, and there is considerable experimental evidence as to how this might be achieved. For example, attempts should be made to make the message as fluent and familiar as possible, taking advantage of variables like repetition, rhyme and easy readability. Statements that sound familiar, as though they have been heard before, invite less scrutiny than unfamiliar statements. Information processing is an individual thing which brings people feelings of ease or difficulty. Any influence that either helps or gets in the way of easy information processing can have a serious effect on how people judge it and the consequent decisions. Easy processing is the aim. And to do that means thinking through and using evidence about the best form of presentation. If it's easy to read, it seems easy to do.

Lesson 2: Decision Making The most powerful mind-changing influence in decision-making is social-consensus. Social psychologists (and advertisers) have long been aware that people often rely on social-consensus to determine whether something is true or not—if so many people believe it, there's probably something to it. But a consensus can only be built on the base of what people already believe. Psychologically, people are more likely to follow the lead of others like themselves, or of others they would like to be like—the current celebrity 'Hello' culture is an underused force for good.

It has been said that 'Nobody cleans a rented car', the reason being no sense of ownership. The massive global arguments can seem to be way outside the individual's ownership and control, their incomprehensibility being counterproductive. It would be sensible to teach responsibility for the environment on the assumption (justified or not) of climate change already germinating in the public mind.

In Sweden, for example, psychologists asked 621 participants aged from 18 to 75 whether 44 statements about climate change were true or false (Sundblad et al.

2007). The big global facts on climate changes, the causes and the consequences for the weather, sea and glaciers, produced little notable response or concern. But in the more personal health statements here appeared to be a sense of ownership which affected responses. When told, 'It is probable that mortality by lung oedema and heart problems during heat waves in Sweden will increase in the next 50 years', that statement produced the strongest reaction.

Influential experimental work by Kurt Lewin in America showed how people's outlooks and productivity could be changed by understanding them in their life-space (Lewin 1948). He pointed out that individuals make decisions within a group, especially when they share a common goal. He used three groups in a famous experiment. The group that was democratically led, where everyone felt they had a part, motivated its members far more than either the autocratic group where members were told what to do, or the laissez faire group without any leadership. To change behaviour, he concluded, the approach should be persuasive and involving rather than either didactic or no leadership.

Schools, apartment blocks, factories, and other institutions act as social groups. It is these social networks which can make 'green' behaviour seem like the normal thing. But not everyone is altruistic: propaganda without action may simply produce eco-fatigue. People want to see the benefits to themselves. When taxes increased on cigarettes, consumption fell. Other financial incentives, such as tax breaks or rebates for solar panels appear to initiate action. In Germany, for example, there is financial help for solar power, whereas there is no such help in the UK. Solar panels on houses in Germany are growing in number, whereas there are few in the UK. But in London, the use of electric cars is increasing rapidly possibly because they are not subject to the congestion charge to enter the city centre and for them parking is free.

We cannot see or feel the effects on global health from what each of us does, so we have to take it on trust that if we recycle paper it is going to make a positive difference to the world. Yet to keep change moving, individuals need clear positive feedback, at very least a pat on the back. Psychologically, we know that when rewards are immediate they are more valued and effective than when they are a long time away. It is often easier to reward the results of the behaviour, rather than the behaviour itself (Winter and Koger 2003). Using green forms forms of energy should be the cheapest. But whether of money or time, the perceived cost to the individual cannot be higher than they are prepared to give.

Lesson 3: Education Education is an environment set within a greater culture. Children who fail to learn about their role as part of a world perspective, to understand and think about life outside their own lives are intellectually restricted. In fact, there is evidence that when children are better educated they are more intelligent and more understanding they are of the outcomes of human action.

There are two ways to help children become more aware of the world. The first is from the more usual direction, 'top-down', when teachers tell pupils what to think and how to behave. But this kind of didactic instruction ignores social-consensus and so may be rejected or swiftly forgotten by the pupil. The alternative is the slower but more effective 'bottom up' approach where learning and attitude change is a more democratic process, involving teachers and pupils learning and thinking together.

To have the greatest effect, education should be of both kinds. A 'top-down' approach could start with coordinating and expanding on what pupils already know and the ideas they have about it. Alternatively, in the 'bottom-up' approach, teachers could organise workshops in schools to start an involving awareness campaign about the energy use of local facilities. Promoting awareness enables every child to expand knowledge in a meaningful way; knowledge which can be used flexibly and creatively in many situations. It is the original meaning of education—bringing out the best from young minds, rather than attempting to fill empty vessels.

The best education encourages children to develop curiosity, problem-solving attitudes and a true love of learning to last them for life. To be useful and useable, knowledge must be gained in a way that is meaningful to the child in his or her world. Children will act most positively and creatively when they have enough self-confidence and courage to experiment with what they know and understand.

Writing from Australia, Volk (2008) says that gifted students, more than others, show interest in the future of the world, in that they want to take action for global interdependence. She sees the gifted as "potential future leaders". I certainly found that in my own research (three decades of follow up on the sample described above), that the gifted they were indeed more interested in world events and had much stronger opinions than the average ability youngster, but their outlooks also correlated very highly with those of their parents and their socio-economic status (Freeman 2010). Briefly, the more intellectual the home, the more the children in it would be involved in thought and consideration of non-domestic happenings. For sure, the gifted have a greater potential to deal with issues of change and morality, though this does not mean that they will certainly take up these matters. I argue that to have their greatest positive effect, global concerns should be a matter for all young people.

Developing a concern for the environment in school pupils is essentially concerned with intercultural understanding and collaboration with regard for cultural viewpoints. It involves, of course, the use of natural resources and what each individual can do for our joint benefit, but also includes concern for peace, international trade, poverty and the availability of clean water and medicine. In much of the developed world such matters are more frequently becoming part of school curriculum from the start.

International communication about the environment is affected by the following:

- *Language and literacy.* English is the primary technological language, but in all languages, literacy is the sure route to flexible thinking and openness to change (Freeman 2008).
- *Cultural approaches.* For example, if one culture sees interference with what they see as natural and interference with God's will, they will refuse to seek change. This could be, for example, a refusal to limit the number of children in a family.

- *Geography*. This includes not only the home area but also distance. Being directly involved with the home district is more effective than secondary information
  about places a long way away.
- *Technology*. This is a two-edged sword. It can be used for good in raising awareness of climate change, or ill in coordinating terrorist activities. The areas of the
  world which have access to information technology are already greatly further
  ahead in communication than those who do not have it.

Lesson 4: Government Legislation It is not only ordinary people who need convincing, but more importantly—politicians. The final stage in the smoking ban
came through legislation. The law is the final decider. But even so, the idea that
smoking in public places is wrong could not have become fact as smoothly as it did
without the considerable backing of social-consensus.

Legislation has brought immense changes for the benefit of the environment,
though this could be speeded further with penalties for polluters, as in the taxes on
cigarettes. Forward-thinking legislators have taken brave steps, such as banning
smoking in Irish pubs, and now banning free flimsy plastic bags in China. Even
greater effects could come from obliging car manufacturers to modify engines and
use more environmentally friendly fuel. Legislation also implies monitoring and
evaluating its possible effects. But it can only function well if there is a basis of
consensus, whether conscious or unconscious. Legislation is perhaps the ultimate
psychological action.

References

Cassidy T (1997) Environmental psychology: behaviour and experience in context. Psychology
    Press, London
Cooper J (2007) Cognitive dissonance: fifty years of a classic theory. Sage, London
Festinger L (1951) A theory of cognitive dissonance. Tavistock, London
Freeman J (1991) Young people's attitudes to nuclear war. Int J Adolesc Youth 2:237–243
Freeman J (2008) Literacy, flexible thinking and underachievement. In: Montgomery D (ed) Gift-
    ed, talented and able achievers. Wiley, Chichester
Freeman J (2010) Gifted lives: what happens when gifted children grow up. Routledge, Brighton
Freud A (1937) The ego and the mechanisms of defense. Hogarth, London
Goldacre B (2008) Bad science. Fourth Estate, London
Hogarth RM (2001) Educating intuition. University of Chicago Press, Chicago. http://www.statis-
    tics.gov.uk/CCI/nugget.asp?ID=828&Pos=3&ColRank=2&Rank=100000
Jung CG (1964) Approaching the unconscious. In: Jung CG (ed) Man and his symbols. Aldus
    Books, London
Lewin K (1948) In: Lewin GD (ed) Resolving social conflicts; selected papers on group dynamics.
    Harper & Row, New York
MacKay DJC (2009) Sustainable energy—without the hot air. UIT, Cambridge. http://www.with-
    outhotair.com/download.html
Nickerson RS (2002) Psychology & environmental change. Lawrence Erlbaum Associates, New
    York
Office for National Statistics report Living in Britain (2008) Retrieved 25 Jan 2008

Sundblad E-L, Biel A, Garling T (2007) Cognitive and affective risk judgements related to climate change. J Environ Psych 97–106. http://dx.doi.org/10.1016/j.jenvp.2007.01.003
Vigotsky LS (1978) Mind in society. The development of higher psychological processes. MIT Press, Cambridge
Volk V (2008) A global village is a small world. Roeper Rev 30:39–44
Winter DD, Koger SM (2003) The psychology of environmental problems. Lawrence Erlbaum Associates, New York

Chapter 8
Promoting Critical Thinking to High School Students When Teaching About Climate Change Through a Participatory Approach

Laura Barraza and Barbara Bodenhorn

Abstract New paradigms in science education are focused on moving towards a sustainable society, meaning redefining the educational practices and developing new methods in order to establish better relationships among individuals, groups, and the society. Being able to reflect upon developing new pedagogic strategies, that support collective action, is crucial to favour social change. Education in the twenty-first century should be based on critical and social theories of the environment and development, in order to link the prospects for sustainability to new forms of economy, social welfare, governance and education (Barraza et al., Environ Educ Res 9(3):347–357, 2003). The nature of contemporary knowledge and knowledge construction demands increasing collaboration and communication between once isolated disciplines. Curriculum integration can reduce curriculum fragmentation, promoting a better awareness of the way different forms of knowledge work and contribute to collaborative knowledge construction, stimulating a critical and a reflexive perspective in their learners. This chapter will focus on the pedagogic strategies used in a research project aiming to provide potential young scientists from rural communities of Mexico and Alaska with a unique opportunity to learn more about their own local knowledge whilst gaining a better understanding of how it intersects with global processes. The project has helped students make cognitive links between their scientific knowledge and life experience, and has established affective and behavioral links which have intensified the ways in which they value their environment, culture, traditions and communities (Tytler et al. 2010; Bodenhorn, Learning about environmental research in a context of climate change: an international scholastic interchange (pilot project). Final report. BASC (Barrow Arctic Science Consortium)). The conjunction of collaborative, interdisciplinary work and multiple pedagogic strategies applied in this specific educational practice has shown the potential of implementing research group initiatives in science

L. Barraza (✉)
Faculty of Arts and Education, Deakin University, Melbourne, Australia
e-mail: laura.barraza@deakin.edu.au

B. Bodenhorn
University of Cambridge, Cambridge, UK
e-mail: bb106@pem.cam.ac.uk

A. Mendonca et al. (eds.), *Natural Resources, Sustainability and Humanity*,
DOI 10.1007/978-94-007-1321-5_8, © Springer Science+Business Media Dordrecht 2012

education. We believe that educational approaches that create "spaces" for students to work together towards a goal defined as a common good, can contribute significantly to develop effective science programs in schools.

8.1   Introduction

One of the main challenges that education faces today deals with the improvement of critic participation: how to engage individuals to become active participants? How can we promote active participation when teachers at schools are not teaching students to act critically? How can we expect to have a sustainable society if we have philosophical, methodological and ontological tensions between the principles of sustainability and the teaching practice at schools? The challenge is to raise global education levels without creating an ever-growing demand for resources (e.g. material, human, economic, etc) that instead of enhancing the quality of education could limit it. For this, it is also relevant to promote a critical, responsible and committed society, but where should we start? Communicating and teaching about sustainability it's a first and fundamental step. Sustainability has emerged as one of the key principles which guide the development process. It is concerned with how best to meet social and economic development objectives without compromising the future viability of natural and human systems (Brundtland 1987).

Meeting these challenges depends on reorienting curricula to address the need for more-sustainable production and consumption patterns. Principles of sustainability reflect upon the urgent need to change the teaching practice at schools. We believed that school teachers need to revise their methodological strategies when delivering information to students in order to enhance their active role and critical thinking, particularly when teaching science and environmental issues. Education for sustainable development (ESD) is based on the belief that everyone has the right to learn, the capacity to contribute and to be committed to global issues to ensure that others can share the benefits of development (Barraza et al. 2003). From a constructivist perspective by reflecting upon experience (Campbell and Tytler 2007), individuals construct their own understanding of the world they live in, learning then the process of adjusting mental models to accommodate new experiences. In doing so, individuals establish links and connections with their understanding. This mental and cognitive activity requires a pedagogical approach that supports collective action and reflection. So we need to learn how to change the way of teaching, from being transmissible to becoming transformative. Besides skills and knowledge, values need to be shared, as do best practices.

Because education for sustainable development is multidisciplinary and interdisciplinary, a holistic approach is essential. The relationship between education and sustainable development is complex. Research shows that basic education is crucial to a nation's ability to develop and achieve sustainability targets. Indeed education can improve agricultural productivity, by learning and applying new techniques or methods; enhance the status of women, by giving them opportunities to find better jobs in different areas of their development; reduce population growth rates, by informing women and the society in general; enhance environmental protection, and raise the standard of living. But the relationship is not linear. Society needs to be aware of the complexity of the world in which one lives and to have the knowledge, critical thinking skills, moral and environmental values and capacity to participate in decision making about environmental and developmental issues. Education for sustainable development is relevant to all types, levels and settings of education. It also constitutes a comprehensive approach to quality teaching and learning since ESD engages educational community with global key issues as human rights, poverty reduction, sustainable livelihoods, climate change, gender equity, corporate social responsibility and protection of indigenous cultures in an integral way (Sterling 2001). But ESD it is not just any kind of education it is about *learning for change* and about *learning to change*. In particular, it is about the content and processes of education that will help us to learn how to live together sustainably (IUCN 1991). For example, one of the most important aspects of teaching, especially in science and environmental education, is that we want our students to be able to relate the concepts they have learned to real world experiences. In order to achieve this we need to ask ourselves some of the following questions: which are the values we teach when teaching sustainability issues? Which cognitive skills do we want to develop when teaching sustainability issues? Which are the teaching skills we need to have in order to promote an attitude that meets sustainability? Some of the values we can transmit when teaching sustainability issues are: to care about something larger than themselves; to be more responsible; to adequate their behaviours to what they believe in; altruism, holistic approaches, open-mindedness, understanding consequences; a care of duty, the importance of community and community spirit; compassion, empathy, sensitivity, selflessness, among others. Some of the cognitive skills we develop are: reasoning, critical thinking, problem-solving, analytical skills; thinking as a sequence, time management and to be able to rationalize actions; to express their own opinion while understanding the validity of other's, and the ability to debate; ability to express themselves in both writing and speech. Teaching skills to students when teaching about sustainability issues are fundamental to build on their character and personality (Watson et al. 2009). Some of these skills are: the ability to communicate clearly and concisely; to be open-minded; to encourage discussion and debate; to encourage questioning and deep thought; to engage students in activities that allow them to show what they have learnt but in an informal and constructive ways; to encourage students to develop their own conclusions, to encourage questions that involve in-depth thinking; to be passionate and motivating. These skills ideally will also help them to develop a critical sense of reflection.

In order to achieve the principles for building a sustainable society, it is fundamental to **develop a pedagogic approach based on values education as a teaching and learning practice**. One way of doing this is through the development of curriculum integration. Curriculum integration can be described as an approach to teaching and learning at all educational levels, that is based on both philosophy and practicality (Primary Programs Framework 2007). It can generally be defined as a curriculum approach that purposefully draws together knowledge, skills, attitudes and values from within or across subject areas to develop a more powerful understanding of key ideas. Curriculum integration occurs when components of the curriculum are connected and related in meaningful ways by both the students and teachers. Sustainability and environmental topics are excellent examples to relate for curriculum integration, because they deal with economical, political, social and health issues. Curriculum integration can reduce curriculum fragmentation, promoting a better appreciation or a sense of awareness of the way different forms of knowledge work and contribute to collaborative and interconnect knowledge construction stimulating a critical and a reflexive perspective in their learners.

The benefits of an integrative approach to curriculum and planning according to the Canadian Primary Programs Framework (2007) are:

- Allowing for flexibility: Teachers can plan for the development of key skills and understandings that transcend individual strands and subjects.
- Building on prior knowledge and experiences: Choosing meaningful connections among subject areas helps students build on their diverse prior knowledge and experiences, it also enables students to forge linkages between taught materials and makes it easier for them to learn and apply concepts and helps them supports their holistic view of the world and ensures more meaningful learning.
- Unifying the students' learning: Students develop a unified view of the curriculum to broaden the context of their learning beyond single subject areas.
- Reflecting the real world: When curriculum is organized in a holistic way, it better reflects the real world and the way children learn at home and in the community.
- Matching the way students think: Teaching ideas holistically, rather than in fragmented pieces, better reflects how young students' brains process information.

8.2   The Practice of Teaching Science

Science and environmental education at school has been taught following traditional methods, that usually do not necessarily respond to students' apprehensions or fail in challenge them to be interested, curious and participative. This has had ontological and methodological implications for the learning and understanding of scientific concepts. For example, the interest of Mexican youth in issues of climate change and local environmental conditions is not accompanied by an understanding about the links between these processes (Barraza and Ruiz-Mallén 2008). We believed that the experience we report in this chapter will contribute to build a new practice in the teaching of sustainability and environmental education involving high school students and all community members.

Place-based education is a pedagogy that science educators are exploring in order to connect students to their community and local place (Tytler et al. 2010). Place-based education lies in the way that it serves to strengthen students' connections to others and to the place in which they live (Tytler et al. 2010). Place-based education provides opportunity beyond classroom walls. It is intergenerational, multidisci-

Fig. 8.1 Mexican and Alaskan students participating in the project

plinary and experiential, uses knowledge and skills in real life situations, is authentically connected to student life-worlds and builds a sense of ecological relationship (Edwards 2003; Smith 2002a). Place-based education encourages ethical action and social responsibility through dealing with community issues. Making connections with the places people inhabit is an important component of well-being and a dimension of a transdisciplinary approach to teaching and learning (Smith 2002b). Following place-based education, a constructivist approach within a sociocultural perspective and the development of multiple pedagogic strategies (discussion groups, field trips, data collection, observations, school-based classes, group and individual presentations, writing essays, among others), we have organized a comparative project involving two Mexican communities and one Eskimo community in the coast north of Alaska (Fig. 8.1). These three communities share the following factors:

- Indigenous communities;
- Communal organization;
- Each community has the responsibility and the right to develop a natural resources management plan;
- Similar political position.

The goals of the project were to provide potential young scientists with an unique learning opportunity to think beyond their local knowledge in a more informed and creative way. To create opportunities for information exchange, communication and education among adolescents, to improve their role as social stakeholders in their communities, to establish international links between young members of indigenous communities with responsibilities for maintaining environmentally sound development strategies. In addition, we aimed to foster a greater sense of appreciation in the young participants for the special qualities of their own communities (Bodenhorn 2007). The program integrated 12 young people (16–18 years old) for a month in a range of scientific research and cultural activities currently being conducted on the North Slope of Alaska (Barrow), and in the mega bio-diverse regions of the Sierra Norte of Oaxaca and the Pátzcuaro region of Michoacán, Mexico. The project was thus holistic, encouraging understanding that emerges at the conjunction of scientific research, experiential observation, language, and cultural processes of various sorts. Pedagogical activities in all three communities included (1) classroom instruction and discussion led by local scientists and project personnel; (2) lab

Fig. 8.2 Students monitoring and taking data using specialized equipment

Fig. 8.3 Teacher in Barrow, Alaska showing students about computer design

and field-based practical engagement with data gathering and processing (Fig. 8.2); (3) an interactive technology (IT) component (Fig. 8.3); (4) engagement with non-science trained local experts in environmental knowledge, and (5) an introduction to local history, language and culture with a specific view to learning more about the bases of local environmental understanding (Fig. 8.4). Students were weekly presented with research questions that were designed to encourage them to draw across the range of information they were receiving, for instance, 'what is an ice calving event' and 'how might one affect Iñupiaq social life'? These questions had to be answered in presentations made at the end of every week as a result of group work. More comprehensive presentations were made before a public audience at the end of every stay in each community. These presentations not only encouraged students to improve their skills in the organization and presentation of complex material, but they also encouraged students to think ecologically, that is, to recognize the inter- connections between factors that include physical, social, economic and political processes. In addition, students individually were required to keep journals, to produce one report on their Barrow research experience and two further essays at the end of the 2 month period (Bodenhorn 2007).

By asking questions, designing investigations, investigating, formulating explanations, presenting findings and reflecting on findings, students were able to explore their own ideas, to compare alternative explanations, to test and evaluate their ideas

Fig. 8.4 Students learning about cultural practices in Barrow

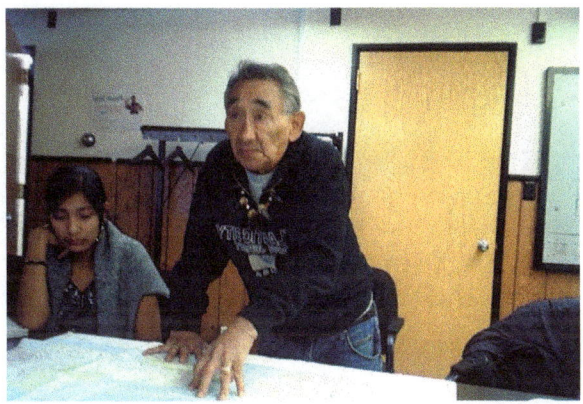

and reasoning with ideas and evidence. Their curiosity and involvement in scientific inquiry moved them beyond passive learning to higher order thinking. The learning and understanding of scientific processes was fundamental by providing the students with the experience of being in the Arctic, and experienced themselves the dramatic evidence of climate change processes that are found in the Arctic.

8.3 Addressing Climate Change in the Arctic

Teaching about climate change presents special challenges to educators. The topic is complex because the Earth's systems are complex, and scientists themselves are not at all certain of the potential ramifications of our interference with these systems (Grant and Littlejohn 2001). Many topics that are relevant when addressing the climate change issue are already part of most curricula: these include, for example, technology topics such as energy systems; social studies topics such as political decision-making; or geography and science topics such as weather systems, photosynthesis and decomposition, and adaptations of plants and animals to specific habitats and climatic conditions. Climate change also has more severe implications than just feeling hotter summers or warmer winters. The participation of the students in this project showed that when exposing to a diversity of pedagogic strategies they were able to enhance their learning and understanding of scientific concepts, establishing links and connections with real examples in their own communities.

During 8 weeks (four in the Arctic and four in Mexico), students exercised their critical thinking when discussing environmental issues; they collaborated in research studies on climate change; they carried out field trips and participated in data collection and monitoring for research studies on Tundra ecosystem and archaeological excavations; they participated in forest management activities and studies on biodiversity in the forest regions of Oaxaca and Michoacan in Mexico (Fig. 8.5); they received lessons on how to apply scientific methods, and participated in discussions on environmental changes; they carried out discussions and exercises in themes such as history, traditional knowledge, and the richness and

Fig. 8.5 Participants in Mexico learning about different techniques of management forests

Fig. 8.6 Participants in Mexico doing some forestry management practices

diversity of natural resources in their communities; and furthermore adolescents learnt about the indigenous culture, history, environment, and language of the Ixtlecos, San Juaneros and Iñupiaq communities (Fig. 8.6). Teaching about and taking action on climate change may not be as difficult as it seems. Teaching should always start with an understanding of what students currently know, rather than with assumptions of they should know, this can be done by doing a diagnostic assessment. Discussion is a powerful tool in having students sharing their ideas (enhancing learning) and provides the teacher with an insight into student thinking. By providing ideas which link to students' prior understanding, you will be giving them additional experiences from which to form their concepts and mental models. A key successful element of participation in this project was the discussion level that students had (Fig. 8.7). A good example of how students were establishing links between their activities and the environments was the discussion they built on Robert Suydam's lecture on Arctic's ecosystems, with a complementary lecture on forest ecosystems and a day long tour through the several ecosystems that form Ixtlán's communal territory. Suydam's discussion on the relationship between non-biological and biological components was crystal clear about why a system as simple as the arctic would immediately reflect climate changes. This was brought home in a different way when students were confronted not only with the mega biodiversity of the Sierra Juárez, but with the maintenance of that biodiversity as part of the community's development goals. Indeed, when the Ixtlán students were prepar-

Fig. 8.7 Students presenting their work

ing their presentation of Ixtlán's natural resources, they found themselves thinking about their own resources in new, more ecological, ways (Bodenhorn 2007). In the previous example shown, students were integrating ideas. Integration is related to enabling students to forge linkages between taught materials what would make easier for them to learn and apply concepts. One of our goals in this project was to "skill students" to use critical and reflective thinking about their own and others' perspectives on a subject. This was successfully achieved due to the different pedagogic strategies used along the project. Some of the learning experiences that students had after this exchange program are expressed in their own voice as follows: "*at school, in my biology class I feel more confident. The knowledge I gained from what I learned about Biocomplexity in the Arctic has helped me established links and to have a better understanding of science… (Atzin)*". "*I have been able to apply some of the things I learnt in Alaska in various ways, with my family I listened to them with respect, at school, I can understand more about cause and effect processes in science… (Rodo)*".

8.4 Future Perspectives

Taking a sociocultural perspective and based on Vygotsky (1986), science teaching involves three fundamental steps:

1. The teacher must make the scientific ideas available on the social plane of the classroom.
2. The teacher needs to assist students in making sense of, and internalizing, those ideas.
3. The teacher needs to support students in applying the scientific ideas, while gradually handing over to the students' responsibility for their use.

These three steps were fundamental in the process of allowing students exploring their own ideas, internalize them and apply them in real life situations. Learning science is about participating in scientific community practices (Campbell and Tytler 2007). This project, together with the school community of three different regions, provided authentic science practices to students allowing them to build an understanding of scientific processes, by connecting, linking, reasoning and establishing comparisons.

This international environmental education exchange program has helped students not only to make cognitive links, between their scientific knowledge and their real life experience, but also to establish affective and behavioral links in order to treasure their environment, their culture and traditions and become better citizens.

The United Nations Environment Program (UNEP) in its Climate Change subprogram is working with countries to strengthen their ability to adapt to climate changes, move towards low-carbon societies, improve understanding of climate science and raise public awareness of the Earth changing climate. One of the goals of this subprogram is "Enhancing knowledge and communication". The experience we have developed in the Arctic, involving indigenous youth from three rural communities, has contributed to improve understanding of climate change science and raise awareness of climate change impacts among decision-makers and other target audiences. Some lessons and ideas of this experienced could be used in a normal school setting by working closely with the learning community (Teachers, students, parents and community members).

References

Barraza L, Ruiz-Mallén I (2008) Estrategia participativa de investigación educativa socio ambiental con jóvenes de una comunidad forestal mexicana. In: Curiel A (ed) Investigación Socioambiental: Paradigmas en salud ambiental y educación ambiental. Universidad de Guadalajara y Universidad de Granada, Guadalajara

Barraza L, Duque-Aristizábal AM, Rebolledo G (2003) Environmental education: from policy to practice. Environ Educ Res 9(3):347–357

Bodenhorn B (2007) Learning about environmental research in a context of climate change: an international scholastic interchange (pilot project). Final report. BASC (Barrow Arctic Science Consortium)

Bodenhorn B, Barraza L, Ruiz-Mallen I (forthcoming manuscript) Toward an ethnography of collaboration. The Roots of Success: a case example

Brundtland HC (1987) BrundtlandReport. World Commisionon Environment and Development, Tokio, Japan

Campbell C, Tytler R (2007) Students' science conceptions, and views of learning. In: Venville G, Dawson V (eds) The art of teaching primary science. Allen and Unwin, Perth, pp 23–42

Edwards J (2003) Butterfly business: connecting science and service learning. Investigating 19(1):8–11

Grant T, Littlejohn G (eds) (2001) Teaching about climate change: cool schools tackle global warming. Green Teacher, Toronto. ISBN: 086571-437-1

IUCN/UNEP/WWF (1991) Caring for the earth: a strategy for a sustainable living. Earthscan, Gland

Primary Programs Framework (2007) Curriculum integration: making connections. Alberta Education, Alberta

Smith G (2002a) Going local. Educational Leadership. Association for Supervision and Curriculum Development 60(1):30–33

Smith G (2002b) Place-based education: learning to be where we are. Phi Delta Kappan 83(8):584–593

Sterling S (2001) Sustainable education: re-visioning learning and change. Green Books for The Schumacher Society, Totnes

Tytler R, Barraza L, Paige K (2010) Values in science, environmental education and teacher education. In: Teacher education and values pedagogy: a student wellbeing approach. David Barlow, NSW, pp. 156–178

Vygotsky LS (1986) Thought and language. MIT Press, Cambridge

Watson P, Wienand N, Workman G (2009) Teaching sustainability: a valid methodology for addressing the associated problematic issues. World Acad Sci Eng Technol 53:908–912

Chapter 9
Participatory Development of Fotonovelas for Environmental Education and Community Empowerment

Anthony Michael Marzolla, A. Yau, H. Grossman and S. L. Dashiell

Abstract Water quality is a serious concern for both environmental and human health. Water quality throughout the world is threatened by human activity, such as unregulated runoff of harmful substances into the watersheds. In Santa Barbara, California, the Latino community, which makes up 40% of the population, was the target of an environmental education initiative that utilized *fotonovelas*, an image-based informational booklet for semi-literate populations, to increase awareness of water quality issues in the community. The fotonovela project was developed to incorporate educational research into environmental education in a participatory manner that involved multiple stakeholders including the Santa Barbara County Water Agency, Project Clean Water, and the local Latino community. The project, coordinated by a team from the Agua Pura program of the University of California Cooperative Extension 4-H Youth Development Program (UCCE), utilized critical pedagogy to determine the water quality concerns of the community and design the fotonovela's educational message. Over 2,000 copies of the fotonovela were distributed throughout the Latino community. This chapter explores why educational research should be used in environmental education, specifically critical pedagogy which aims to empower underserved or oppressed constituents of society through the process of analyzing, naming, and characterizing issues in the form of dialogue. The fotonovela project demonstrates how environmental education initiatives guided by critical pedagogy framework and based in participatory development can be successful in changing behaviors that affect water quality in Santa Barbara.

A. M. Marzolla (✉)
Cooperative Extension 4-H Youth Development Program, University of California, 1265 Ferrelo Road, 93103 Santa Barbara, CA, USA
e-mail: ammarzolla@ucdavis.edu

A. Yau · S. L. Dashiell
Bren School of Environmental Science & Management, University of California, Santa Barbara, CA, USA

H. Grossman
Department of Education, University of California, Santa Barbara, CA, USA

A. Mendonca et al. (eds.), *Natural Resources, Sustainability and Humanity*,
DOI 10.1007/978-94-007-1321-5_9, © Springer Science+Business Media Dordrecht 2012

9.1 Introduction

When we speak of "horticulture" or "agriculture," we speak of the word "culture" as meaning "to cultivate, to grow." Michael Cole, an education researcher at UCSD (2007), suggests that when we speak of human culture we are talking about a similar process, the process of cultivating humans to develop in particular patterns or ways. A plant can grow and produce in the wild, uncultivated by humans, but nonetheless is still shaped by environmental conditions. When we intentionally cultivate a plant, we must be concerned with its growing conditions—not just conditions such as soil type in the immediate vicinity, but also with humidity, temperature, sunlight requirements, and pollutants found in the general region. Similarly, when we cultivate a human's growth, we cannot be concerned only with the conditions in our learning environment, but must address conditions in the overall society as well.

This metaphor of culture as a garden changes culture from a static to a dynamic concept. Culture is not just seen as the process of helping things grow, but is converted into a general theory for how to promote growth (Cole 1995). Currently we cultivate students in a manner analogous to industrial farming: we educate them in a homogenous manner broken down into specific cultivating patterns. "Skill practices" replace "watering," "lessons" replace "fertilization," and the "school year" replaces "growing cycles". Unfortunately, our school system is most analogous to a mono-cropping system that treats all of our diverse students as if they were a single variety of plant. Whereas the ideal learning conditions for each student might be substantially different. Historically, when looking at learning differences, researchers have been tempted to attribute individuals successes and failures to genetics. The work of Ogbu and Simon (1998) used comparative research to conclude that no cultural group is more successful in school because of genetic superiority. Rather, their research suggests that it is the interaction between the student and their learning environment that seems to be an indicator of a student's success. When the learning conditions fit the student, the student will flourish.

While the school is one place to address these growing concerns and difficulties, it is not the only place. One such arena for exploration is extra-curricular environmental education programs. The field of environmental education is unique in that, for the most part, environmental educators keep their dialogue separate from the dialogue of the rest of the field of education. Perhaps this is in part because having an instructor with adequate and accurate environmental knowledge is often seen as more important than having an instructor with educational expertise. In fact, a simple web search will show some entry level positions in the field of environmental education that do not include any educational experience as a job requirement. However, this is not to say that the environmental education field is at odds with the general field of education, or that it could not gain from expanding its dialogues to include more educational research. Throughout this chapter the role of critical pedagogy in environmental education will be explored: why it is necessary, how it ties to the goals of environmental education, and how we apply it in an environmental education context to develop a participatory educational initiative relating to water quality.

9.2   The Basics of Critical Pedagogy

To use educational research in the field of environmental education, one must begin with some sort of theoretical framework in which to move. This can be a daunting task given the range and scope of educational research. In searching for a paradigm, we began with our goal—to examine the role of water quality in our community—and looked for a framework with which this goal could be best met. Plants have difficulty communicating their needs. As humans, we use language to communicate our thoughts, wants, and desires. Through the process of communication and dialog we can reach a greater awareness of the world around us and what we determine is important. "Critical pedagogy" is a theoretical framework based on this increased awareness through dialog. Critical Pedagogy stems from the work of Paulo Friere, a Brazilian educator who changed the education field with his book, "Pedagogy of the Oppressed," published first in Portuguese, in 1970 (Friere 1970), then translated into English. In the work of Friere and other Critical Pedagogists, education is grounded in the concept of conscientizacion—this is the idea that by cultivating a critical awareness of life and its conditions, institutions and social groups can alter the power structures that create these conditions and change the conditions themselves (Duran 2011). These goals are achieved through a process of "praxis", a reflective process of analyzing, naming, and characterizing issues in dialogic form. Through examining the underlying conditions that shape current situations, participants from all walks of life are able to analyze the issues that are important to address and determine how to best address them. This empowers communities to make the changes most important to them. Fundamental to this pedagogy is the belief that communities must transform their relationships with the population at large to not just tolerate the diverse beliefs, values, and linguistic tools of a community, but to respect them. This can be a challenge when differences are considered deficits.

9.3   Culture as Politically Charged

Unfortunately, culture is not a politically neutral subject. Rather, "culture is deeply entangled with economic and political privilege" (Nieto 2010, p. 141). Pierre Bourdieu (1986) suggests thinking of this in terms of "cultural capital." In his work, Bourdieu looks at three kinds of capital that people hold within a community: economic capital, social capital, and cultural capital. Economic capital can be seen as our monetary system; social capital can be seen as the social networks that connect to economic capital; and cultural capital can be seen as the values and tastes that tie you to a privileged social and cultural class. In our society, certain cultural practices are given more weight simply because they are associated with the dominant social class. If students' practices are not in line with the dominant cultural practices, they are often seen as arriving with a cultural deficit (Bourdieu 1986).

In order to negate the underlying idea that groups can be culturally deficit, we must begin by understanding the cultural contexts with which people enter their educational paths. The discussions that bring these concepts to light are not always easy or conflict-free, however; they respond to people as individuals instead of as groups and attempt to draw this discussion from a deficit perspective to an additive perspective. Educational practices that use culture as the starting point for education are known as culturally relevant pedagogy. This branch of education is founded on the belief that students' backgrounds are assets that students can draw upon in their own learning, assets that should be nurtured and developed instead of ignored and devalued. Culturally relevant pedagogy can be particularly powerful when combined with a critical pedagogy to create an approach that teaches students to work through ideas and to collaborate together. Students are asked to struggle through discussions of difficult ideas and social justice concepts knowing that conflict will arise. In fact, in this pedagogy, the conflicting views are necessary to explore the changing face of culture and to turn learners into actors involved in shaping their own realities (Nieto 2010).

9.4   The Role of Language in Cultural Capital

An important aspect of cultural capital that critical pedagogy seeks to address is language. Much of our society still equates use of the primary language in a culture with intelligence. To illustrate this, we take an example from *The Literacies of Power*, by Donald Macedo (1994). Macedo immigrated to the United States during his high school years. As an intelligent critical-thinker, he was determined to go to college. However, when he spoke of his goals with his school's college counselor, the counselor suggested that instead it was best that he become a T.V. repairman due to his lack of English speaking skills. Macedo was already a scholar who spoke three languages fluently and had scored highly in both French and Spanish achievement tests. His counselor couldn't see past his limited English proficiency, seemingly unable to address him and his goals from anything except a deficit perspective. The languages he spoke were not seen as carrying much cultural capital, and his intellectual worth was degraded in the counselor's eyes. Luckily for the field of research, Macedo did not let this discourage him from his educational goals and he has added greatly to the dialog on critical theory.

Even if Macedo's experience were an isolated one, it would be disappointing. Unfortunately, stories similar to his are repeated on a regular basis so much so that these inequalities became the center of his research. When addressing issues of language, Macedo does not want the discussion to be whether English should be the only language used in school, an argument that he sees as placing English as a superior language with more cultural capital. Instead, Macedo wants the discussion reframed in terms of the ideological goals of different forms of language use that confront the racism, classism, and economic disparities embedded in our educational system: a critical pedagogic approach to language. Author David Corson agrees with this assessment, writing about the inherent problems with English-Only education, "It avoids the fact that decision makers cannot see the world from the point of view of those who are very different from themselves and who do not enjoy the same privileged language position" (2001, p. 29). In this approach, speaking multiple languages would be seen as additive instead of creating a deficit that needs to be rectified. When viewed in this light, studies show that additive bilingualism enhanced students' cognitive, linguistic, and academic growth (Cummins 1996). Thus, we argue that a bilingual approach should not be avoided in education. Rather, the curriculum should utilize as many cultural and linguistic resources as possible to improve their education impact.

9.5   Relevance to Environmental Education

One might question how readily culturally relevant pedagogy might be applied to environmental education. Science education, more than most, tends to think of itself as culturally neutral—though, as mentioned above, culturally relevant pedagogy would argue that this is not the case with any subject. How then do the culturally active goals of critical pedagogy fit the goals of environmental education?

The Environmental Protection Agency states its education goals as follows:

> Environmental education enhances critical thinking, problem-solving, and effective decision-making skills, and teaches individuals to weigh various sides of an environmental issue to make informed and responsible decisions. Environmental education does not advocate a particular viewpoint or course of action. (EPA 2010)

Culturally responsive critical pedagogy nurtures these same thinking skills. Critical thinking is developed through the negotiations necessary in the dialogic processes. Problem-solving skills and decision-making skills are addressed through collaborative work within a group to identify, classify, and analyze issues the community itself identifies. Also, just as environmental education should not advocate for a particular viewpoint or course of action, critical pedagogy is about empowering and hearing multiple voices, not just one. If these stated goals of the EPA are in fact what our society wants from environmental education, then a culturally relevant, critical pedagogy is ideal for studying environmental issues embedded in diverse cultural settings.

9.6   Education in the Latino Community

For the first time in modern California history, Latinos are now the majority constituents of our schools (CDE website 2010). Of a total of 6,191,110 students enrolled in California schools, 3,118,717 of them are classified as Hispanic or Latino of any race. That takes their percentage to 50.4%, up 1.36% from last year. In Santa Barbara County, where our organization conducts its work, 41,433 of 65,960 students are classified as Hispanic or Latino. That is 62.8% of the County's school population. Due to historical and structural inequalities, ranging from culturally deficit perspectives to English-only classrooms, there is an education achievement gap between Latino students and others that still needs bridging. In Santa Barbara County Latino students are twice as likely to drop out of school than are white students (CDE website 2010).

How do we address the educational needs of the Latino population? The first step comes with recognizing that the traditional education techniques we have been using with these populations have not been fully successful. The next step involves working within their communities to determine how these learning goals are best met. To do this, we use insight from critical pedagogy and culturally responsive pedagogy to determine the needs of these communities and collaborate to address these needs.

9.7 Methods

9.7.1 The Agua Pura Program

Agua Pura is a program originally developed in 1999 as a partnership between the University of Wisconsin's Environmental Resources Center, Santa Barbara City College, and the Santa Barbara County Cooperative Extension 4-H Youth Development Program. The results of this collaboration are described in detail in our online manual for those interested (Andrews and Marzolla 2001). Its goal is to work with the Santa Barbara Latino community to develop and adapt resources and strategies for watershed educational needs and interests. It does this through a variety of activities, including youth leadership training, interactive workshops, and curriculum development. In 2007 we received a grant from the Santa Barbara Water Agency to assist the agency in complying with NPDES (Non-Point Discharge Environmental Standards) under the Clean Water Act and to develop best practices in reaching under-served audiences. The goal was to use empowerment models to involve the Latino community in the development of promotional materials to be used in the community context. We decided to apply critical pedagogy techniques in the formation of this promotional material. The following will be an account of our project and the benefits of using a critical pedagogic approach in environmental education.

9.7.2 The Participatory Process

We began this process by looking to the local Latino community for partners in this development. We found two: the Latino families involved with the Tri-Counties Resource center in Santa Barbara and *promotoras* who met at the Isla Vista Youth Center. *Promotoras* are peer educators that reach out to underserved communities as liaisons between their communities and health and social organizations. They

Fig. 9.1 Promotoras with program staff at the Isla Vista Youth Center pose for a photo after a survey session

conduct their work as members of the communities with which they liaise and are often quite successful because they speak the same languages and share life experiences with the populations they serve. Promatoras are generally women whose community roles include advocate, educator, mentor, outreach worker, role model, and translator. Their methods vary in different areas (Proyecto Vision Website 2010). Here, in Santa Barbara, the promatoras meet regularly at the Isla Vista Youth Center and could be addressed as a group (Fig. 9.1). It should be noted here that there is a particular struggle in the Santa Barbara area to get members of the Latino community to participate in organized activities around environmental topics. This is attributed by participants to fear, lack of transportation, lack of childcare, and feelings of disenfranchisement. Despite these concerns, the Agua Pura team was able to reach 64 participants between the information gathering and feedback stages of our project.

9.7.3   Our Meetings

After we had embedded our research within the community where our actions would be developed, we held a series of community dialogue meetings. In these meetings we surveyed community members about their knowledge of water quality issues, assessed perceptions of local water quality, and discussed practices in and around the home that might be affecting water quality. One commonality that arose from these meetings was that about 90% of our participants did not know about basic water quality issues, including tap water safety and proper disposal of household chemicals. Their lack of knowledge did not, however, seem related to lack of concern, as the majority of participants had significant personal experiences with water quality issues and held concerns for their children in regards to water quality. Another disturbing trend was that none of the participants were able to clearly identify resources in the community for water quality programs and information. They were very vocal about the need for more access to information regarding these resources and their accessibility to the community.

9.7.4 Surveys

A survey was designed to gather information from the community regarding the level of awareness regarding municipal water resources and water quality. The survey included questions relating to perceptions about local water quality, practices in the home that affect water quality, and the use and/or enjoyment of water resources. There were 17 questions, written both in English and Spanish to avoid any misinterpretations due to language barriers and were administered by bilingual surveyors. These interviews were conducted individually with community members so they would feel more comfortable giving complete answers to each of the questions and results were anonymous. The administrators of the survey were given a sheet of survey questions complete with answers and further information. They used these responses to guide further discussions, intentionally bringing up areas where community members expressed concern to assure that critical pedagogy was being practiced. Discussions were often initiated in response to survey questions and in this way the surveys became a significant part of the educational initiative.

9.7.5 Identifying the Project

After surveying our community groups and collecting the information from our discussions, the Agua Pura team went about using the information from these collection methods to create promotional materials to be used in the community context. A fotonovela was chosen as the primary material for disseminating information on water quality. The story depicted by pictures would focus on the water quality issues that could most readily be addressed in the house, as identified by the survey. These included the proper disposal of motor oil and household chemicals, and water conservation.

9.7.6 The Background of Fotonovelas

Fotonovelas, a form of graphic novel, originated as melodramatic love stories, and target semi-literate women, although men read them as well (Flora and Flora 1978; Carrillo and Lyson 1983). This form of media is popular throughout Latin America (Flora and Flora 1978). In the United States, the fotonovela is popular in the Chicano/Latino community, often addressing important social issues within the community, and has been suggested to serve as a cultural bridge for American Latinas between an unfamiliar society and their traditional one (Carrillo and Lyson 1983). Fotonovelas have been viewed negatively as tools for reinforcement of traditional class values and consumerism (Flora and Flora 1978), but, in general, they are considered to encourage increased literacy (Emme et al. 2006) and empowerment (Wang and Burris 1994).

The use of fotonovelas has since been adapted for outreach and education, most often for public health education (Wang and Burris 1994; Velarde 2002; Clark et al. 2004; Ross 2004; Valle et al. 2006). They have also been applied in occupational and environmental health education (Peres et al. 2006), and as a tool for immigrant youth to explore the challenges of attending school in a different culture (Emme et al. 2006; Kirova and Emme 2006; Kirova et al. 2006). In the latter context, the photography done through the fotonovela process allowed the children and researchers who spoke different languages to communicate (Kirova and Emme 2006; Kirova et al. 2006).

9.7.7 Creating the Storyboard

Our student interns—also members of the community we were attempting to include in our outreach—used the survey feedback to design a storyboard that addressed the gaps in the community's awareness of water quality issues (Fig. 9.2). They created a few storyboards with different versions of the story we wanted to tell about proper waste disposal, keeping water clean, and the drinkability of tap water in the area. These storyboards were taken back to the community for vetting. The most successful story was then made into a Spanish language fotonovela using local community members as actors and community resources for filming, printing, and distribution. These techniques paid close attention to cultural practices and were forms of culturally-relevant pedagogy.

The final story, "Carlos and Clean Water" or "Carlos y el Agua Limpia" told the story of how a teenager named Carlos helped his family with the chores around the house while educating his family on water issues (Figs. 9.3 and 9.4). The first lesson that he taught related to how to clean outside areas, such as driveways and sidewalks. In the next lesson, Carlos helped his dad figure out proper disposal of paint and motor oil. This is followed by a mini-lesson on trash disposal in regards the watershed system. Finally, in the last section, the drinkability of tap water is covered.

Fig. 9.2 Program staff working on the draft fotonovela storyboard

The pictures, the language, and the storyboard were derived from the community to which the story was targeted. For this reason, the fotonovela was well-received (Fig. 9.5). In addition to the storyboard, the fotonovela booklet provided detailed information on the resources available in their own neighborhoods for properly disposing of waste, learning more about water quality, and engaging with water quality issues. The fotonovela was the attractive medium by which this information was transmitted.

9.8   Results

Over the next 2 years 2,000 copies of the fotonovela were distributed at community events and through the work of the promotoras, who valued the fotonovela so much that they took it to community members during their outreach work. They perceived the use of these fotonovelas as a straightforward way to convey important information, and felt that one of the benefits of the fotonovela was that it was an honest representation of daily life and related activities.

9.9   Discussion

When we began this project, one of our goals was to use educational research, specifically critical pedagogy, in an environmental education model to determine how successful it would be in an out-of-the-classroom setting. The results we obtained from our project led us to believe that this method can be successfully implemented in such a manner. Not only were our fotonovelas well received by the community at large, but the process of creating the fotonovelas was educating in and of itself.

Fig. 9.3 "Carlos y el Agua Limpia" ("Carlos and Clean Water"), page 2 of the fotonovela, Carlos educates his family about wasting water

Fig. 9.4 "Carlos y el Agua Limpia" ("Carlos and Clean Water"), page 8 of the fotonovela, father tells his son to make sure he disposes of his trash properly

Fig. 9.5 "Carlos y el Agua Limpia" ("Carlos and Clean Water"), cover page of the fotonovela with an outline of the booklet's content

First, we, as an organization intent on educating the Latino population about water quality issues, gained significant insight into community needs and awareness through our informal group discussions. Just as importantly, these dialogs in and of themselves increased interest in and enthusiasm for issues of water quality amongst community constituents. This is evidenced by the promotoras interest in sharing the fotonovela with their outreach work. The feedback from most participants was positive, to such an extent that the promotoras requested more education and training on water quality information and discussed the idea of making a series of fotonovelas on similar issues.

Overall, we would term the project as being very successful and are currently working on another fotonovela project, this one involving school-aged children creating luchador (masked wrestlers popular in Mexico) characters. We do not hesitate to suggest that other environmental education programs attempt to use critical pedagogy in their educational practices. These techniques created powerful conversations that brought environmental issues to the forefront of the minds of communities typically hard to reach by institutional groups and empowered community members to become actors involved in changing practices in their area. Furthermore, we urge other institutions to reconsider English only use in their programs where applicable. A great deal of our success came from our willingness to use a community's native language to address community concerns. We must constantly be aware that our educational practices are not done in a neutral environment, rather they are shaped by cultural, linguistic, and political concerns, no matter what the setting. By addressing these issues in regards to any topic being discussed, our work is more likely to have a positive outcome.

References

Andrews E, Marzolla A (2001) Agua Pura: a leadership planning manual for U.S. Latino communities. Environmental Resources Center, University of Wisconsin, Stevens Point

Bourdieu P (1986) The forms of capital. In: Richardson JG (ed) Handbook of theory and research for the sociology of education. Greenwood Press, Westport, pp 241–248

California Department of Education (2010) Data Quest (data file). Available from http://dq.cde.ca.gov/dataquest/. Accessed April 2010

Carrillo L, Lyson TA (1983) The fotonovela as a cultural bridge for Hispanic women in the United-States. J Pop Cult 17:59–64

Clark R, Baron M, Molina G, Alvarado M, Teran L (2004) What my girlfriend didn't know: developing a low-literacy bilingual fotonovela on folic acid. Genet Med 6:332

Cole M (1995) Culture and cognitive development: from cross-cultural research to creating systems of cultural mediation. Cult Psychol 1:25–54

Cole M (2007) Phylogeny and cultural history in ontogeny. J Physiol Paris 101(4–6):236–246

Corson D (2001) Language diversity and education. Lawrence Erlbaum Associates, Mahwah, pp 16–35

Cummins J (1996) Negotiating identities: education for empowerment in a diverse society. California Association for Bilingual Education, Los Angeles, pp 1–368

Duran R (2011) Development of Latino family school engagement program in us contexts: enhancemets to cultural historical activity theory accounts (in press)

Emme MJ, Kirova A, Cambre C (2006) Fotonovela and collaborative storytelling: researching the spaces between image, text, and body. Prairie Metropolis Centre, Edmonton, pp 1–8

Environmental Protection Agency (2010) Environmental education background and history. http://www.epa.gov/enviroed/eedefined.html. Accessed April 2010

Flora CB, Flora JL (1978) The fotonovela as a tool for class and cultural domination. Lat Am Perspect 5:134–150

Friere P (1970) Pedagogy of the opressed. Seabury, NY

Kirova A, Emme MJ (2006) Using photography as a means of phenomenological seeing: "doing phenomenology" with immigrant children. Indo Pac J Phenomenol 6:1–12

Kirova A, Mohamed F, Emme M (2006) Learning the ropes, resisting the rules: immigrant children's representation of the lunchtime routine through fotonovela. J Can Assoc Curric Stud 4(1):73–96

Macedo D (1994) Literacies of power. Westview Press, Boulder

Nieto S (2010) Language, culture, and teaching: critical perspectives. Routledge, NY, pp 1–280

Ogbu JU, Simon HD (1998) Voluntary and involuntary minorities: a cultural-ecological theory of school performance with some implications for education. Anthropol Educ Q 29(2):155–188

Peres F, Moreira JC, Rodrigues KM, Claudio L (2006) Risk perception and communication regarding pesticide use in rural work: a case study in Rio de Janeiro State, Brazil. Int J Occup Environ Health 12:400–407

Proyecto Vision website (2010) http://www.proyectovision.net/. Accessed April 2010

Ross FC (2004) Developing and implementing culturally competent evaluation: a discussion of multicultural validity in two HIV prevention programs for Latinos. New Dir Eval 102:51–65

Valle R, Yamada AM, Matiella AC (2006) Fotonovelas: a health literacy tool for educating Latino older adults about dementia. Clin Gerontol 30:71–88

Velarde LA (2002) Minority adolescents as health educators: personal perspectives American alliance for health, physical education, recreation & dance national convention and exposition. San Diego, California

Wang C, Burris MA (1994) Empowerment through photo novella: portraits of participation. Health Educ Behav 21:171–186

Chapter 10
Participation and the Construction of Sustainable Societies

From Rhetorical and Passive Participation to Emancipative Democratic Dynamics

Mário Freitas

Abstract There are very different meanings associated to the word *participation* and the concept of *participation* is used in very different situations and backgrounds, with very different objectives: (a) from very passive and/or manipulative senses (people are only informed or very indirectly eared throw the opinion emitted by their elected representatives); (b) to more extended, powerful and autonomic senses (that promote real conditions of participation both for individuals and communities and guarantee a real impact of individual/communitarian opinions in the decisions taken).

Based on a deep theoretical literature review and the analysis of Brazilian and Portuguese realities we argue that the generality of participatory processes related to territorial planning and development options going on in both countries can be classified in the lower levels of hierarchic participation models (manipulative or passive participation). In some cases we can identify the intention of matching intermediate levels (like functional participation), even though almost always in a very formal and incipient way of implementation. Higher levels (interactive, self-mobilized and autonomic models) are not frequent.

We will present our own conception of participation as a cultural issue and discuss it in the perspective of our spider model of sustainable development. We will defend the necessity of deeper forms of participation to construct more sustainable societies and different types of development. We will finish presenting the main dilemmas, constraints and/or stimuli for a real emancipatory participation.

M. Freitas (✉)
Geography Department, Environmental Studies Centre (NEA),
Santa Catarina State University (UDESC), Florianopolis, Brasil
e-mail: pmariofreitas@gmail.com

A. Mendonca et al. (eds.), *Natural Resources, Sustainability and Humanity,*
DOI 10.1007/978-94-007-1321-5_10, © Springer Science+Business Media Dordrecht 2012

10.1   In the Kaleidoscope of Participation: Terms, Meanings and Levels

The general idea of participation is very old; however the modern debate about participation has only 50 or 60 years. In fact,

> one of the most important roles for public participation emerged in the literature concerning development issues in the late 1950s, when failures of development projects were in many cases attributed to inadequacies of project design and implementation, as a consequence of insufficient involvement by local populations. (Rahnema 1992, quoted by Blamey et al. 2000, p. 2)

More recently the inclusion of personal contributions of those who are affected by development projects become not only an effort to increase project success, but also as an ethical necessity related to the empowerment of communities, so that they are able to participate more conscientiously in decisions which affect them. Following this kind of approach some authors (Slocum and Thomas-Slatyer 1995, quoted by Blamey et al. 2000, p. 2) consider that participation as two main objectives: (a) to increase the probability of success of project implementation; (b) to satisfy the right that all stakeholders have in being involved in the decisions that affect themselves.

Collins and Ison (2006, s/p) stresses that "Participation has become a key consideration in the discourses and practices of environmental policy-making at local through to international levels". However they also point out that "although the imperative for participation and stakeholder involvement has increased the desire and capacity for critically engaging with notions and epistemologies of participation has perhaps lagged". And this is really important because before dealing with other aspects, the discussion about participation, participation objectives and participation intentionality is needed and it implies the discussion of ideologies, cosmovisions and their ontological, epistemological and ethical bases.

Some of the most well-known researchers and academics agree in recognizing the existence of different terms, meanings, levels and assumptions related to the concept of *participation* (Arnstein 1969; Warburton 1997; Uemura 1999; Finger-Stich and Finger 2003; Maranhão and Teixeira 2006; Kouplevatskaya-Yunusova 2006; Bowen 2008). Finger-Stich and Finger (2003, p. 23) goes further sustaining that "one of the reasons this 'politically correct' term is so popular" is because "it can serve as so many purposes". Kouplevatskaya-Yunusova (2006) consider that the controversial nature of the term participation "help to explain the wide range of attitudes of various stakeholders" and conclude that "any type of participatory process is ruled by a hidden agenda, because the ability to influence a decision brings power" (p. 489).

10.1.1   Different Umbrella Terms

The discussion about those different meanings, levels and assumptions of the participation concept begin with the umbrella-term that is utilized. Some authors use several times the umbrella-term participation (Mannigel 2002; Finger-Stich and Finger 2003; McCall 2004); however sometimes they also use umbrella-terms in which an adjective is added to the word participation.

Other authors use predominantly or exclusively composed terms. The term citizen participation[1] used by researcher as Arnstein (1969), Paizano et al. (2006) and Bowen (2008) and is also utilized by several Portuguese speakers' researchers (Maranhão and Teixeira 2006). The expression public participation[2] is used by researchers as James and Blamey (1999), Blamey et al. (2000) and Innes and Booher (2004) is also extensively present in literature of European countries as an umbrella designation that includes very different types of participation. Hesselink et al. (2007) use both the umbrella terms participation and public participation; Innes and Booher (2004) use both public participation and community participation. Some other authors (Uemura 1999; Dreyer 2000; Weiner et al. 2002; Aref 2010) prefer to use the term community participation as a more general term, but some of them (Dreyer 2000; Weiner et al. 2002) consider an explicit connection and identification between it and the expression public participation.

In South American tradition, including in Brazil (that is a Portuguese speaking country), the designations more commonly used as umbrella terms, both in common and academic discourses, is participation[3], but also popular participation[4] (Amorim and Sato 2010) and communitarian participation[5]. Still other umbrella terms are used, like social participation[6] (Costa and Pascoal 2004). In a similar way Toro and Werneck (2004) use the term participation including it in the more general context of social mobilization[7]. Collins and Ison (2006) put the emphasis in social learning and see participation (as well as informing and consultation) as a necessary but not sufficient condition to reach it.

10.1.2 Different Meanings

Warburton (1997, p. 7), quoting Oakley (1991) stresses that, "it is generally recognized that Participation defies any single attempt at definition or interpretation" but add that however "it is crucial that some common understanding of participation is achieved in order to progress the debate". Arnstein (1969) early considers that the attempts of defining the concept of participation have always been integrated in the logic of political contention and so "are purposely buried in innocuous euphemisms like 'self-help' or 'citizen involvement' " (s/p). At the same time, even though ap-

---

[1] Participação cidadã, in Portuguese.

[2] And the correspondent participação pública in Portuguese.

[3] Participação, in Portuguese.

[4] Participação popular, in Portuguese.

[5] Participação comunitária, in Portuguese.

[6] Participação social, in Portuguese.

[7] Mobilização social, in Portuguese.

parently situated in an opposite side, the rhetoric expression 'absolute control' goes in the same sense of emptying the real challenging nature of the process. Finger-Stich and Finger (2003, p. 23) sustains that "if the goal of participation processes is not the growth or survival of state organizations… the concept of participation needs to be clearly defined in each situation".

One of the first modern definitions of participation is the one of Arnstein: "citizen participation is a categorical term for citizen power". Some other general definitions are given by: (a) Buttoud (1999a, b p. 17) for whom *participation* means "to take part of debate or an action"; (b) Weiner et al. (2002) for whom *public participation* "refers to grassroots community engagement" (p. 2); (c) Talbot and Verrinder (2005), quoted by Aref (2010, p. 1), consider *community participation* as an effort to "bring different stakeholders together for problem solving and decision making"; (d) Putnam (2000), quoted by Aref (2010, p. 1), for whom *community participation* "plays an essential and longstanding role in promoting quality of life".

Oakley (1991), quoted by Warburton (1997, p. 7), gives the following contribution to participation definition:

> Participation is concerned with human development and increases people's sense of control over issues which affect their lives, helps them to learn how to plan and implement and, on a broader front, prepares them for participation at regional or even national level. In essence, participation is a 'good thing' because it breaks people's isolation and lays the groundwork for them to have not only a more substantial influence on development, but also a greater independence and control over their lives. (Oakley 1991, p. 17)

Finger-Stich and Finger (2003, p. 23) defending the necessity of defining the term in each situation, for the proposals of his text, defines participation as:

> The voluntary involvement of people who individually or through organized groups deliberate about their respective knowledge, interests, and values while collaboratively defining issues, developing solutions, and taking—or influencing—decisions. (Finger-Stich and Finger 2003, p. 23)

The same author clarifies that, according to this definition "strikes, boycotts, and demonstrations" are not forms of participation because "they do not allow a two-way exchange or open and voluntary deliberation" (id. ibid.) even though they could be effective strategies to reach effective participation.

Considering that participation is a very different thing for different people, Bowen (2008, p. 65) gives this example:

> In a community development context, participation is 'the inclusion of a diverse range of stakeholder contributions in an on-going community development process, from identification of problem areas, to the development, implementation and management of strategic planning' (Schafft and Greenwood 2003, p. 19). In relation to an anti-poverty programme, it means the involvement of local citizens in various aspects of the programme, from planning to evaluation.

Hesselink et al. (2007, p. 215) defend that public participation is "based on democratic principles" and represents "an approach used by governments, organisations and communities around the world to improve their decisions (…) and involve people who are affected by those decisions". For Toro and Werneck (2004) participation in a social mobilization process is an objective to be reached and a way for reaching other objectives. Finger-Stich and Finger (2003, p. 23) sustains that in order to distinguish those different meanings and uses of the participation concept we should consider the answers to some central questions: "Who participates? How does one participate? About what decisions and issues does one participate? Why does one participate? When does one participate?" Considering that the majority of literature classifications insist more on the first four points (who, how, about, why), the same author claims the centrality of the fifth question (when).

Some authors (Blackmore 2006; Collins and Ison 2006), as it has already been referred, consider that participation is a necessary but not sufficient condition to reach *social learning*. We can still find other conceptions like the one of Innes and Booher (2004, p. 429) that conceive participation "a multi-way interaction in which citizens and other players work and talk in formal and informal ways to influence action in public arena before it is virtually a foregone conclusion".

So we can find different definitions of participation (varying from some very general to some other more specific and context dependent). Oakley (1991, p. 270), quoted by Warburton (1997, p. 7), defend that "even though there seems to be a consensus on paper about what participation is", there are "clearly fundamentally opposed views as to what participation means in practice". In fact, the majority of the researchers recognize the existence of practices that represent very weak and passive forms of participation (non-participation or *soft participation*) and more strong, effective and significant understandings of the concept (*hard participation*).

Various authors also consider the tension between state (and/or powerful economic or political *stakeholders*) control and people or citizen control (Arnstein 1969; Finger-Stich and Finger 2003). Mészáros (2010) goes further and states that true participation implies a deep change of our capitalistic system, including the political framework throw which capitalism today dominantly operates, the parliamentary system. For Mészáros (2010, p. 16) true participation in a new social-economic-politic order will be "the completely autonomous self-management of the society by producers freely associated in all domains and far beyond of restricted mediations (obviously still necessary during some time) of the modern political state" (Mészáros 2010, p. 16). But it is impossible to discuss *participation* meanings without discussing levels or degrees of participation.

10.1.3   Levels, Degrees or Types of Participation

The framework (Fig. 10.1) for considering different types, levels and/or degrees of participation is closely related with the discussion we have done before.

Arnstein (1969) stresses that the presentation of her "typology of eight levels of participation may help in analysis of this confused issue" (meanings of participation). As we will refer in point 1.3.2., other authors (Paul 1988, quoted by Dreyer 2000; Shaeffer 1994, quoted by Uemura 1999; Mannigel 2002; adapted from Borrini and Feyerabend 1996; Pimbert and Pretty 1997; Aref 2010; adapted from Lek-

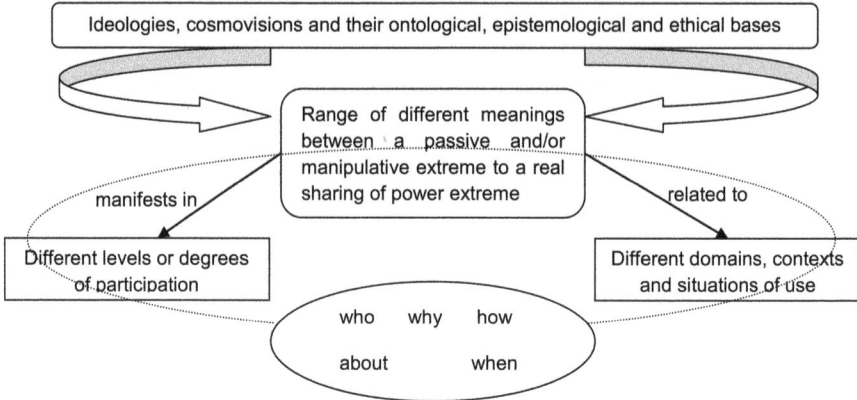

Fig. 10.1 Framework for considering different levels domains or degrees of participation

sakundilok 2006; Aref et al. 2009) present categorizations that can be largely compared with Arnstein's ladder (Table 10.1).

However some other authors (Tritter and McCallum 2006; Collins and Ison 2006) criticize Arnstein's hierarchical categorization based in the idea that participation is sharing power:

> The lack of complexity in the conceptualization of the protagonists in Arnstein's model, its failure to consider the process as well as outcome, or the importance of methods and feedback systems, means that a more nuanced model is required to guide current challenges to user involvement and public participation. (Tritter and McCallum 2006, p. 158)

In any case we will begin by revising the classical approach of Arnstein and other related systematizations. Later, when presenting our own vision of participation we will return to this particular issue and discuss in which way we can combine Arnstein's approach with more dynamic visions of participatory processes.

10.1.3.1 Arnstein's Ladder of Citizen Engagement

Arnstein's Ladder of Citizen Engagement (Arnstein 1969) represents the first relevant systematization (in a modern sense) of different levels or degrees of participation and emerged from Arnstein's work in urban planning (USA in the 1960s). Arnstein's ladder includes the following steps: *manipulation, therapy, informing, consultation, placation, partnership, delegated power* and *citizen control*.

The lowest level of participation ladder (really non-participation) is *manipulation*.

> In the name of citizen participation, people are placed on rubberstamp advisory committees or advisory boards for the express purpose of "educating" them or engineering their support. Instead of genuine citizen participation, the bottom rung of the ladder signifies the distortion of participation into a public relations vehicle by powerholders. (Arnstein 1969, sp)

Table 10.1 Comparison or degrees/levels/types of participation following different authors

Mannigel (2002)[a]	Levels of community participation (Paul 1988)[b]	Arnstein's ladder of citizen participation (Arnstein 1969)	Aref (2010)[c]	Hesselink et al. (2007)	Shaeffer (1994)[d]
6. Transferring authority and responsibility		8. Citizen control	6. Empowerment	6. Self-mobilization	Participation "in real decision making at every stage"
		7. Delegated power			Participation as implementers of delegated powers
5. Sharing authority and responsibility in a formal way	4. Initiating action	6. Partnership	5. Partnership	5. Functional participation	Participation in the delivery of a service, often as a partner
4. Negotiation and developing specific agreements	3. Decision-making	5. Placation	4. Interaction	4. Participation for material incentives	Involvement through the contribution (or extraction) of money, materials, and labor
3. Seeking consensus					
2. Active consulting	2. Consultation	4. Consultation	3. Consultation	3. Participation by consultation	Involvement through consultation on a particular issue
1. Informing	1. Information sharing	3. Informing	2. Informing	2. Passive participation	Involvement through 'attendance', implying passive acceptance of decisions
		2. Therapy 1. Manipulation	1. Manipulation	1. Manipulative participation	Involvement through the mere use of a service

[a] Adapted from Borrini and Feyerabend (1996) and Pimbert and Pretty (1997)
[b] Quoted by Dreyer (2001)
[c] Adapted from Leksakundilok (2006) and Aref et al. (2009)
[d] Quoted by Uemura (1999)

Some forms of manipulative participation are very well known because (even though mixed with other forms), they have been and still are dominant in the last 40 years. Several meetings or committees in which people are invited to participate but which the main proposal is only or predominantly advising, persuading or even educating and in which the role of citizens is to hear and or sign a proposal are manipulative forms of non-participation.

As Arnstein (1969) claims, based on a poster painted and used by French students during the events of May 1968[8] "participation without redistribution of power is an empty and frustrating process for the powerless". It allows the powerholders to claim that all sides were considered, but makes it possible for only some of those sides to benefit. "It maintains the *status quo*" (Arnstein 1969, sp). So, from Arnstein's point of view, the top of her ladder is exactly *citizen power*.

> It is the redistribution of power that enables the have-not citizens, presently excluded from the political and economic processes, to be deliberately included in the future. It is the strategy by which the have-nots join in determining how information is shared, goals and policies are set, tax resources are allocated, programs are operated, and benefits like contracts and patronage are parceled out. In short, it is the means by which they can induce significant social reform which enables them to share in the benefits of the affluent society. (Arnstein 1969, sp)

But for Arnstein *citizen power* is not some kind of give all power to all people; in fact, no one has the absolute control of all situations and dynamics. When we talk about citizen power we are talking about a certain degree or control that guarantees that people "can govern a program or an institution, be in full charge of policy and managerial aspects, and be able to negotiate the conditions under which "outsiders" may change them" (Arnstein 1969, sp).

Between the two extreme steps of Arnstein's categorization there are more six levels. Beginning by the lowest level immediately after manipulation, we have *therapy* (also a non-participation process in Arnstein's opinion). Therapy mechanism can be of different kinds but some of the most common are therapy groups for adjusting tenant behavior related to community rubbish or prevent aggressive behaviors, etc. For Arnstein, therapies are used to substitute genuine participation and "their real objective is not to enable people to participate in planning or conducting programs, but to enable powerholders to 'educate' or to 'cure' the participants" (Arnstein 1969, sp).

Informing, consulting and *placation* are successively, from down to top, the next three steps of Arnstein's ladder; they are tokenistic forms of participation or "levels of minimum concession of power" (Amorim and Sato 2010), that

> … allow the have-nots to hear and to have a voice: (3) Informing and (4) Consultation. When they are proffered by powerholders as the total extent of participation, citizens may indeed hear and be heard. But under these conditions they lack the power to insure that their views will be heeded by the powerful. When participation is restricted to these levels, there is no follow-through, no "muscle," hence no assurance of changing the status quo. Rung (5) Placation is simply a higher level tokenism because the ground rules allow have-nots to advice, but retain for the powerholders the continued right to decide. (Arnstein 1969, sp)

---

[8] O poster tem escrito; "je participe, tu participes, il participe, nous paticipons, vous participez, ils profitent".

We all know very well different kinds of informative campaigns. This kind of approach is generally based in one-way communication processes implemented by media, pamphlets/folders distribution, poster display, or inquiries. Meetings, conferences and public sessions of some institutions (parliaments, local power meetings, etc.) are also very divulgated forms of *informing*, several times presented as participation. Informing can be an important part or participatory process but alone is not real participation.

Consultation consists basically in inviting citizens to give their opinions. This can also be an important step towards a full participation but if it is "not combined with other modes of participation, this rung of the ladder is still a sham since it offers no assurance that citizen concerns and ideas will be taken into account" (Arnstein 1969, sp). The most frequent strategies used for *consultation* are attitude surveys, neighborhood meetings, and public hearings. "Attitude surveys are not very valid indicators of community opinion when used without other input from citizens" (Arnstein 1969, sp) because as the same author argues *survey after survey* studies demonstrated that sometimes people really don't know what and how much they can suggest and so only suggest very instrumental and simple things.

Consultation meetings if not included in a more organized process of full participation can be demobilizing and frustrating and lead people to passive attitudes.

Public hearings has been politically and legally promoted as top instruments of participation in territorial planning and management, as well as in the discussion of projects of development and studies of environmental impact; however a large amount of public hearings (if not preceded by other mechanisms of participation or only preceded by restrict specialized or strict representative meetings) are poor forms of participation and several times became manipulative processes and they really doesn't have great influence in decision making. Agra Filho (2010, p. 355), referring to the requirements for public hearings about studies of environmental impacts stresses the importance of: (a) previous analysis of "the quality and methodological pertinence and sufficiency of data"; (b) "previous divulgation in a accessible speech of the reports"; (c) "adequate procedures for enabling an effective discussion of different issues"; (d) "meeting dynamic oriented for the validity and objective of the project, impact preview and mitigation measures of those impacts". The same author concludes that

> … the available studies and data as well as the observations resulting from my experience of participating in those hearings show that in the majority of public hearings it has been recurrent the precariousness or even neglected of those requirements. (Id. Ibid.)

In Arnstein's opinion the real sharing of power begin with the level of *partnership*.

> At this rung of the ladder, power is in fact redistributed through negotiation between citizens and powerholders. They agree to share planning and decision-making responsibilities through such structures as joint policy boards, planning committees and mechanisms for resolving impasses. After the groundrules have been established through some form of give-and-take, they are not subject to unilateral change. (Arnstein 1969, sp)

A more extended and deep form of sharing power before reaching the more elevated level of the scale (*citizen power*) is *delegated power*.

> Negotiations between citizens and public officials can also result in citizens achieving dominant decision-making authority over a particular plan or program. (…). At this level, the ladder has been scaled to the point where citizens hold the significant cards to assure accountability of the program to them. (Arnstein 1969, sp)

Arnstein's ladder is really very considered and referred. Indeed in almost all the articles that make an adequate analysis about participation meanings and levels contain more extended or restricted references to that classic categorization. However as we have already referred (and we will discuss more in a next point) several critics have been formulated from different authors.

10.1.3.2   Other Categorizations

Even though some of the authors don't stress the hierarchical feature of the categorizations, they really include an idea of hierarchy, from non-participation, very passive or soft participation to more powerful, hard and emancipatory kinds of participation. Several of those characterizations can be related and compared (Table 10.1).

In order to discuss the roles of community participation in the development of education Aref (2010, p. 2) talks about different "types of community participation for educational planning and development" and specifically distinguishes between:

- "empowerment" when "local people have control over all development without any influence (Choguill 1996; Dewar 1999)";
- "partnership" if "there are some degrees of local influence in development process (Arnstein 1969)";
- "interaction" if "people have greater involvement" and "the rights of local people are recognized and accepted in practice at local level" (Pretty 1995);
- "consultation" when "people are consulted in several ways" like "community's meeting or even public hearings" and "developers may accept some contribution from the locals that benefits their project" (Arnstein 1969);
- "informing" when "people are told about development program, which have been decided already, in the community" and "the developers run the projects without listening to local people's opinions" (Arnstein 1969).
- "manipulation" if development is promoted by "some powerful individuals, or government, without any discussion with the people" (Arnstein 1969).

As we can easily conclude this approach is fairly based in the classic Arnstein's ladder.

Shaeffer (1994), quoted by Uemura (1999), distinguish between the following levels (except perhaps the levels we marked with an * can largely matched with Arnstein's ladder[9]):

---

[9] That items can in any way be compared with one of the levels presented by Hesselink et al. (2007) referred to below.

- involvement through the mere use of a service (such as enrolling children in school or using a primary health care facility);
- involvement through the contribution (or extraction) of money, materials, and labor;
- involvement through 'attendance' (e.g. at parents' meetings at school), implying passive acceptance of decisions made by others;
- involvement through consultation on a particular issue;
- participation in the delivery of a service, often as a partner with other actors;
- participation as implementers of delegated powers;
- and participation "in real decision making at every stage" including identification of problems, the study of feasibility, planning, implementation, and evaluation.

Hesselink et al. (2007), regarding people and community involvement in biodiversity conservation and management present a seven level categorization that also can be (except perhaps in what respects to "participation for material incentives") largely matched with Arnstein's ladder:

- "Manipulative participation"—participation is simply pretense, with "people's" representatives on official boards but who are unelected and have no power;
- "Passive participation"—people participate by being told what has been decided or has already happened (information); external agents define problems and information gathering processes and so control analysis; it doesn't concede any share in decision making, and professionals have no obligation to take on board people's views;
- "Participation by consultation"—people participate by being consulted or by answering questions; external agents define problems and information gathering processes, and so control analysis; such a consultative process does not also concede any share in decision making;
- "Participation for material incentive"—people participate by contributing resources, for example, labor, in return for food, cash or other material incentives.
- "Functional participation"—participation seen by external agencies as a means to achieve project goals (… reduced costs). People may participate by forming groups to meet predetermined objectives related to the project; such involvement may be interactive and involve shared decision making, but tends to arise only after major decisions have already been made by external agents;
- "Interactive participation"—people participate in joint analysis, development of action plans and formation or strengthening of local institutions; participation is seen as a right, not just the means to achieve project goals; the process involves interdisciplinary methodologies that seek multiple perspectives and make use of systemic and structured learning processes.
- "Self-mobilization"—people participate by taking initiatives independently of external institutions to change systems, they develop contacts with external institutions for resources and technical advice they need, but retain control over how resources are used; self-mobilization can spread if governments and NGOs provide an enabling framework of support; such self-initiated mobilization may or may not challenge existing distributions of wealth and power.

10.2   Reframing Participation

As it has already been referred there are not enough assessment studies about citizen participation in the majority of domains of public politics in different countries and especially in Portugal and Brazil. So it's difficult to sustain general conclusions. Anyway there is wide evidence in the literature about deficiencies in participation processes as well as suggestions for optimizing it.

10.2.1   Desilusions and Dilemmas of Participation

Agrawal and Gibson (1999), Campbell and Vainio-Mattila (2003), as well as Friedman (1993, 1998), Hagg (1999), Leach et al. (1997) and Michener (1998), all quoted by Dreyer (2000) refers that participation doesn't meet the expectations in several processes and countries. However, Dreyer (2000) considers that there is "little evidence of a critical assessment of the real need for participation, relevant processes and techniques or their outcomes".

Based on the results of a research workshop they organized Hunsberger and Kenyon also conclude that "on average the participants believe that participatory processes are not very good at representing different values" (Hunsberger and Kenyon 2006) or are a "gloomy picture" (Kenyon and Hunsberger 2006). However, in the same paper they also bring about "that there are good ideas, methods and examples that can be used as lessons in other approaches" (Hunsberger and Kenyon 2006). In the workshop of Path project the participants suggested the importance of: the "institutionalization of public participation in the decision making process"; the "institutionalization of participation within society"; "responding to participatory outputs and increased transparency". In addition, and in what specially respects to the participatory methods: "being clear about participation"; "developing better, more innovative methods"; "better learning from previous participation".

Some of the authors that defend the integration of participation in the general context of social learning point out that social learning can be "understood as one or all of the following":

1. The convergence of goals, criteria and knowledge leading to more accurate mutual expectations and the building of relational capital. If social learning is at work then convergence and relational capital generate agreement on concerted action for integrated catchment management and the sustainable use of water. Social learning may thus result in sustainable resource use.
2. The **process of co-creation** of knowledge, which provides insight into the causes of, and the means required to transform a situation. Social learning is thus an integral part of the make-up of concerted action.
3. The change of behaviors and actions resulting from understanding something through action ('knowing') and leading to concerted action. Social learning is thus an emergent property of the process to transform situations (Slim 2004a; Collins and Ison 2006, sp).

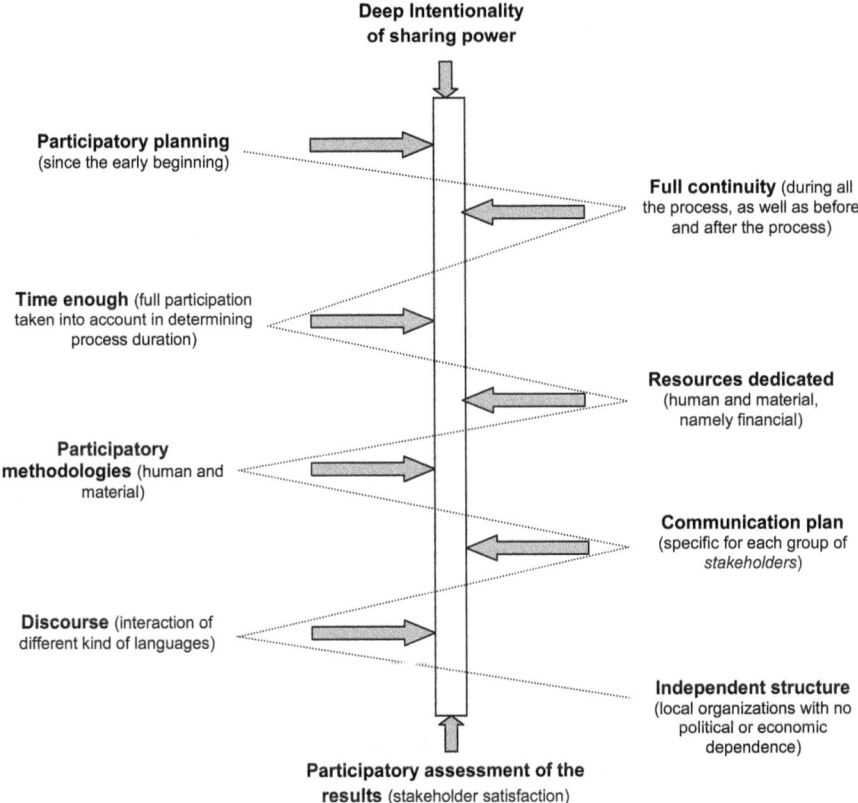

Fig. 10.2 Main constraints/stimuli (and assessment/evaluation criteria) of participatory processes

The same authors refer that the main characteristics of such approach are: "interdependencies", "complexity", "uncertainty" and "controversy".

Based in the revision of relevant literature and on our own experience in participatory processes in Portugal and Brazil we present here what we consider the main constraints (and as at the same time can be the main stimuli) of participatory processes and also assessment criteria of those processes (Fig. 10.2).

As can be seen in Fig. 10.2 we consider that everything begins with the deep intentionality of sharing the power of the decision. That is to say if the institution or agency that is responsible for deciding the public politic or designing the plan or want to develop a project should believe in participation and must have the genuine intention in hearing citizen/community opinion. In the majority of the situations it's not clear this is true. In many cases before hearing citizen/community politicians or managers begin talking and acting as the decision was already done.

This, in a certain way, is happening with the project of Belo Monte Dum, Amazonia, Brazil. There are many evidences that it is some kind of political-economic-developmental obsession that begin with the government of President Fernando Henrique (Brazilian Social Democratic Party) and then passed to the government of

President Lula (Workers' Party) by some kind of damn inheritance. It seems that it was never really put for discussion. Due to the great pressure of indigenous people and scientific community it has been possible to diminish the reservoir from 1,200 to 400 km^2, but... not construct the damn seems to be out of question. It's possible that Belo Monte will be one of the most expensive and less productive plant of Brazil... but it has to be constructed. It doesn't matter if it was one of the possible reasons for Marina Silva abandonment or for the last demission of ICMBio[10] President... or even if it was necessary Ibama[11] to give a partial license that in justice opinion is illegal... etc. A lot of developmental projects promoted by big economic private groups have the same problem. In general the capitalist companies want to have profits and so they are interested in convincing people about positive project impacts and not worried with people opinion.

Another example of this kind of error (with some differences) was the decision of constructing Ota Airport in Portugal. Several attitudes have demonstrated that there was no intention of hearing opinions or sharing the power of the decision. The statements of the Prime Minister at the date defend that all possibilities had been studied and that one was the best. In an interview given to the radio Antena 1 in 12 April 2007 by Socrates (the Prime Minister) declared that "it would be an error" to suspend the project and also emphatically considered (to Antena 1 and RTP 1[12]) that "the country was discussing the matter for 30 years". Suddenly in the middle of great politic pressure and even opinions within the same political sector that things were not so clear, the prime Minister decided to order a study to National Laboratory of Civil Engineering. The conclusion was different and the Portuguese government changes the airport location for Alcochete, a very different geographic location.

Both cases, even though they include other errors (we think they do), are examples of lack of real and deep intentionality of sharing power of decision. It seems the decision was already taken and was not under discussion.

Other important points of constraint or possibilities of promoting deep participation are:

1. the existence or the lack of a participatory planning;
2. the full continuity of the participatory process (it is for example explicitly preview in the Law for the Participatory Municipal Director Plans elaboration in Brazil, but not always followed);
3. enough time (participation, namely where and when there is no previous tradition of participation, should have sufficient time to be prepared and implemented and be integrated in some previous and future plan of full participation);
4. resources dedicated (if money and human resources are not allocated to participatory process, participation is a mirage);

---

[10] Institute Chico Mendes for Biodiversity Conservation, hardly pressed to give a positive opinion.

[11] Brazilian Institute of Environment and Renewable Natural Resources.

[12] Portuguese public television channel.

5. participatory methodologies (making informative sessions with possibilities for making questions and/or giving opinions are not really the best participatory methodologies and even though there are some general methodological resources that can be used in different situations, each concert project or plan shall have a specific methodology planning and implementation);
6. the existence or not of a communication plan with specifications for different *stakeholders* is absolutely central in participation promotion;
7. discourses and languages are the ways in which people, scientists, technicians, politicians, local leaders, etc. express their knowledge, emotions, opinions, etc., and there are many different kinds of discourses and languages that shall be put in interaction in a full process of ecology of knowledge, techniques and discourses;
8. Independent structure (that's to say) real citizenship dynamic with any kind of direct control no political (government or parties) or economic (corporations)
9. finally, the assessment of the results (without a final reflection about what has been decided and in which way the decision took into account all the opinions) is absolutely necessary not only for one specific process of participation but also for future processes.

In several cases, the only kind of participation foreseen on the Director Plans of Brazil and Portugal is the public hearing, and this is not only lacking of planning but also of several other points, like no full continuity since the beginning, no money or human resources allocated, no time enough, no participatory methodologies, no communication plan, among others.

Some more examples from Brazilian reality will be useful to explain a little bit more these constraints/stimuli. The Master Plan of Florianopolis, for example, was at the beginning relatively well organized even though the municipality (it takes too much time, more than a year, to designate the Core Manager Committee[13]), wasn't enough committed with the process and/or didn't give enough support to the process. When the *communitarian* diagnostic was ready the municipality decided to interrupt the work of the Core Manager Committee as well as the *Local Nucleus* saying that the civil participatory dimensions was complete and now is time for technicians to work until the public audience. When the municipality tried to organize a public hearing in April 2010 a big contestation movement organized by social movements with support of the Core Manager Committee and the *Local Nucleus* boycotted the public hearing. The local government tried to approve the project without public hearing but it was unable to face new big popular protests. It is easy to identify what were the main constraints of this process. It begins probably with a non-real desire of sharing power, followed by unilateral rupture of the full continuity of the participatory process and errors in participatory methodologies (namely, creating a pseudo differentiation between public participated phase and technical phase).

Another last example can be presented. It's the case of GERCO[14]. Despite a real concern of coordinators the participatory planning (perhaps resulting, among other

---

[13] Núcleo Gestor.
[14] Coastal Management.

things, from having not enough time and/or money) had some relevant faults. The workshops with sector representatives were only hold in one municipality of each of the five regions and during the day. So it was impossible or very difficult for several representatives of the more distant municipalities to be present. It was a problem of a deficient participatory methodology mixed with short time and resources. But there were other problems. The elaboration or revision of several Director Plans of those coastal municipalities (more than 30) has been done, almost at the same time, by another corporation. The lack of contact between the two planning movements was really very negative and the compatibility of both planning instruments became without resolution. It was a case of lack of planning. At the same time the Director Plans elaboration was an environmental compensation for the enlargement of the road BR 101 but there was not enough money and resources allocated to the process and so almost the 30 Master Plans were made with only an incipient kind of advisory participation (public hearing), even though the Ministry of the Cities defends the participatory director plans and has the guide for that.

An interesting case was that local authorities of some few municipalities, namely Araranguá (in the South) and Governador Celso Ramos (in Center/North) and so integrated in GERCO process. didn't accept the idea of making a Director Plan without participation, pressed the enterprise and/or decided to support and organize the participative process. However, namely in Araranguá (that we have been following with more detail) despite all efforts put through by the coordination of the process, some problems happened in the last phase: technical language used in public hearings was not understood by the people in the meetings. The process is still on course with two public hearings a week...

10.2.2   Understanding in Which World We Live in… Participating

Today's world is dramatically different in what concerns wealth distribution (richness of some few is always but we are not able of finishing with poverty). This wealth differentiation has a great impact in many other differences such those with respect to health care, nutrition, mothers and child mortality, access to potable water, comfort, happiness, etc. It is yet and hopefully different in some cultural aspects, even though the colonial, the post-colonial and the globalization movements produce a deep destruction or seriously affection of cultural diversity (traditional knowledge, traditional artifacts, musical or artistic expressions, etc.).

However, more than in other moments the world today is dramatically unified in what respects to the economic global system: "the majority of the countries became a part of an unique integrated system, the global capitalism" (Reich 2008, p. 2), "super-capitalism" (Reich 2008, p. 5), global capitalist economy (Soros 1999). The central objective of capitalism in general, and of the super-global capitalism (or if we prefer, the neoliberal capitalism) in particular, is the profit: "the objective of the capital in not to satisfy certain necessities, but to produce the profit" (Marx 1970, quoted by Guiddens 2005): "the capital' proprietaries search for maximizing the profits" (Soros 1999); or as some critical defenders of capitalism system say (not explicit referring the question of the profit): "the role of capitalism is to grow the economical cake" (Reich 2008, p. 2).

Some authors (Reich 2008; Soros 1999, 2008) postulate the existence of a substantial distance between economy and society or, if we prefer, capitalism and democracy. For those critical capitalism defenders "the division of the slices and their distribution between private goods, like personal computers, and public goods, like a clean atmosphere, compete to the society. It's the role we assign to democracy" (Reich 2008, p. 2). In the same way goes the opinion of Soros (1999) that refers a certain tendency of considering that democracy and capitalism go on holding hands and assert that the things are more complicated: "capitalism needs democracy as a counterpoint, because capitalistic system on his own doesn't reveal a tendency to equilibrium (…). Acting with total freedom and guided by their own mechanisms they [the capitalists] would go on accumulating capital until the situation attains the disequilibrium" (p. 30).

The ascension of capitalism has strong relationships with the fight against absolutist monarchies, the emergence of liberal revolutions and the consolidation of parliamentary system. However, with the course of time capitalism has supported several dictators and several authoritarian regimes.

Capitalism cannot live without free market, some kind of deregulation, certain degree of unemployment, systematic capital concentration and profit increase, etc., but if it is necessary it can live and/or can be temporarily justified… without liberty or democracy until the moment the dictatorial regimen begins creating problems for capitalistic development.

Until now capitalism has promoted and/or supported military dictatorships (like Chile during the 1970s and 1980s, Egypt), authoritarian monarchies (Saud Arabia, South Yemen, Egypt, etc.) and other kind of authoritarian governments. One of the most dynamic capitalistic countries nowadays, (China) lives with an authoritarian regimen of unique party (auto named as communist).

Reich (2008, p. 1) remembers a very significant episode:

In March 1975, the economist Milton Friedman accepted an invitation of Chile government to have a meeting with Augusto Pinochet, who eighteen months ago had put down the democratic government of Salvador Allende that had achieved power by free elections. Friedman was very criticized by American media but there is no any reason to believe that he approved Pinochet methods. Friedman went to Chile for exhorting the military junta to adopt capitalism of free market—that means to pruner the business regulation and the welfare state that had grown in Chile during several years of democratic regime in the country and open the country to the activities of free trade and investment with the rest of the world. In a series of conferences that he pronounce in the country he confirmed his conviction that free market was a condition necessary for political freedom and sustainable democracy. Pinochet accepted the advice of Chicago economist in what respects to the adoption of free market[15] but his brutal dictatorship prolonged for more fifteen 15 years".

---

[15] Only during some years. The so-called Chilean economic miracle only occurs after the decision of abandoning the recommendations of neoliberal mentors.

And in what respects to Chile we shall remember that some of the main mentors and supporters of the brutal and bloody military punch were the multinational capitalist companies and some conservative national capitalistic sectors.

The history of Bolivia also is an example how far capitalism can go in order of manipulating to have "comprehensive" and "collaborative" regimens.

I think it's not necessary to say nothing more for to conclude that the relationship between capitalism (and specially neoliberalism) and democracy or sustainable society "is in the better hypothesis, tenuous" (Soros 1999, p. 160) or, I would say, only a relationship of utility: democracy is good if it doesn't control to much the growing of profit. "The companies are not moral or immoral" (Reich 2008, p. 10) and the capitalism is not moral or immoral it is amoral (Soros 1999).

> The companies are not citizens. They are mountains of contracts. The objective of companies is to participate in economical game with the higher competitiveness and efficacy. The challenge for us citizens is to avoid that they impose the rules of the game. (Reich 2008, p. 12)

Regarding to the capitalism the importance of participation is almost reduced to market participation. Some passive forms of participation are reduced to basic levels as sharing information or some basic forms of consultation in order of better understand our thinking and feeling as consumers so to optimize our participation in the market game and obviously the profits of capitalists. Other eventual forms of participation, as for example the participation as small shareholders, is only symbolic and manipulative and not real. We receive a letter of the President of the Administrative Council of the Bank informing about an Assembly... but real nothing is going to happen for us. We just don't have any power. The capitalists, the great actionists are going to decide how and when they are going to play with our money in order to enlarge the usually enormous Bank's profits. Environment and environmental impacts are seen by global capitalism as externalities. There are constraints to business or more recently (for example with carbon credits) as an opportunity of business and so an opportunity for making money with the same capitalist rules we have been talking about. Participative processes as the ones foreseen in law related to development projects with environmental impact, are seen by global capitalism as a piece of the game, part of the costs (if they can manage to go ahead with the project) or state impositions, constraints to free market dynamics. In any case the global capitalism, as a system, is not concerned with the epistemological, ethical, moral, social or political aspects of citizenship participation.

So, following the advices of some unsuspected critical defenders of capitalism we shall care about democratic control of capitalism and deposit the hope in democracy, in public power and in the representatives we regularly elected. But, particularly after the 1970s "the capitalism of free market win; but the democracy have lost the freshness" (Reich 2008, p. 1). The pressing exercised by the centers of economic power, inspired by the mentors of Washington Consensus (that is to say neo-liberalism), reduce dramatically in many countries the role of the state and so the possibility of state control.

The debate about economic globalization (that is to say about global capitalism) and its consequences is a debate about economic theories and values (including the value of participating, namely in developmental options and decision-taking) (Stiglitz 2007). As the same author emphasizes 25 years ago, there were three main schools of economic thinking with empirical examples occurring all over the world: free market capitalistic economy, controlled free market economy and "communist" approach[16]. With the fall of Berlin Wall (even though the theoretical school of socialist/communist economy goes on trying to reframe its thinking) we are restricted to the other two approaches. One is dominant nowadays—neoliberal free market theories—and the other goes on trying to reform the dominant way of thinking and acting—controlled free market economy.

In Stiglitz (2007) opinion there is a great difference between these two types of approaches. The "Washington Consensus" gives little importance to equity (...). "Some believe that equity belong to politics' forum and not to economy" (Stiglitz 2007, p. 55). The other school "see the state with a more active role, either in promoting development or in poor's protection" (Stiglitz 2007, p. 56).

> In practice the defenders of this alternative perspective also highlight employment, social justice and non-material values, like environment protection, than the ones that defend the minimalist role of state. (...). It's usual that the proponents of this proposal complain political reforms for giving more voice to citizens in decision-taking process.

So it's not difficult to understand why in a world dominated by global capitalism in which the neo-liberal of free market perspective is dominant we predominantly have manipulation and/or passive participation. Authors like Reich (2008), Soros (1999, 2008) and Stiglitz (2007) believe that it's possible inside the capitalistic system allied with parliamentary democracy.

Other authors like Mészáros (2010), however, criticizing the bureaucratic and autocratic state socialism that fail in Soviet Union and East Europe, still defend the necessity and the inevitability of a new economic and political order and relates this change with the question of *participation*.

> The necessary alternative to parliamentary system is in strict relationship with the question of a true *participation*, defined as self-management completely autonomous of the society by the producers freely associated in all domains, much forwards than the restrict mediations (obviously necessary during some more time) of the modern political State. (Mészáros 2010, p. 16)

10.2.3   Sustainable Societies, Human Sustainable Development and Emancipatory Participation

Based in our sustainable development spider model (adapted from Freitas 2007) and other frameworks formulated in other moments (Freitas 2008; Freitas et al. 2009; Leite et al. 2005), we conceptualize participation in a way that matches some points of view of Innes and Booher (2004).

---

[16] Effectively a certain kind of state socialism controlled economy.

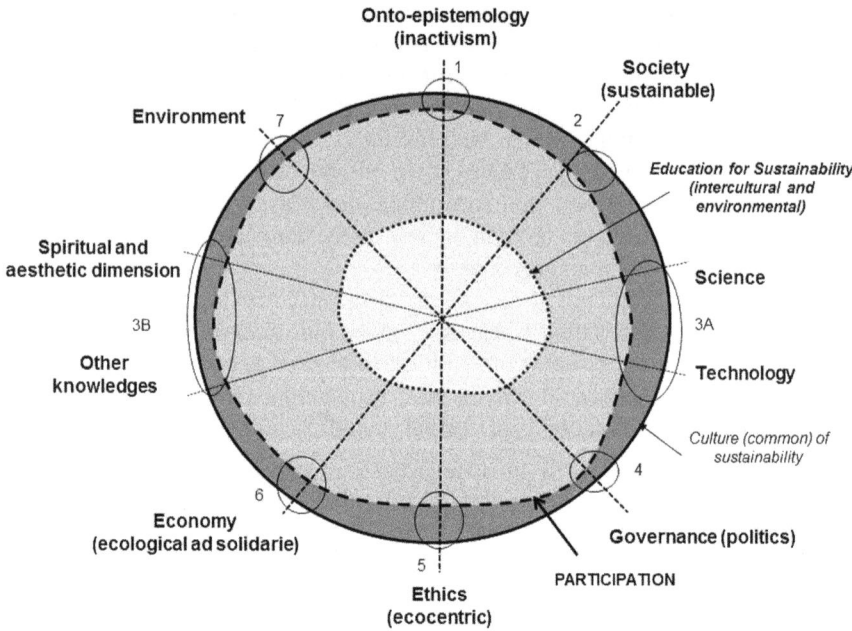

Fig. 10.3 Participation conceptualized as a cultural issue in the context of a new common culture of sustainability as suggested by the spider model of sustainable development. (Adapted from Freitas 2009)

Participation is a cultural (way of living) issue guided for an ideal, a cosmovision, that is not completely defined (especially in what respects to spatial-temporal details) and we even don't know how far and/or when it can be fully attained; so, uncertainty, controversy and complexity are some of the most relevant characteristics of participatory processes. Any way participation should be seen as a network of multipolar interactions guided by a set of new paradigms (represented in Fig. 10.3 by numbers from 1 to 7).

1. It is impossible to separate *to be* (ontological dimension) from *to know* (epistemological dimension) because *to know is to* do and *to do is to know* (Maturana and Varela 2002) and so a participatory process is a *doing-knowing* experience of social learning that shall be so continuous as possible. This approach clarifies the discussion pointed out by Tritter and McCallum (2006) about participation as an objective. Participation is simultaneously a process and an objective and we cannot separate one thing from the other.
2. The construction of more sustainable societies in a global and local point of view (even though is very difficult to say precisely what a sustainable society is) is the goal that shall inspire all the participatory process of decision-taking. And exactly because the construction of more sustainable societies is not a closed predefined end but an open and always on-going process all participatory processes shall have a continuous transformative intend.

3. (A & B) Participation shall be based and promote a real interaction and comprehension between different kinds of knowledge and techniques (scientific knowledge, traditional knowledge, communitarian knowledge, etc.), discourses and languages (scientific, communitarian, scholar, etc.) and also between other inappropriate dichotomies like rationality-emotions and materiality-spirituality. A new scientific and technological is needed

4. The developing of more democratic and emancipatory ways of governances and making politics shall be a goal for participation processes and so, it shall contribute to the reframing of not only all types of clearly authoritarian governance, but also the formal and limited political framework of representative democracies, largely or almost exclusively based in elections ritual. Participation shall develop new democratic approaches as direct democracy, new ways of relating representative and direct democracy, new ways of thinking and implementing decentralizations, new ways of organizing the democratic system (including party's dynamics), new ways of organizing world community (namely ONU), etc.

5. A new ethical approach that substitute the traditional anthropomorphic paradigm by a cosmo-geo-bio based ethical approach that emphasizes the central role of *father-universe* and *mother-earth* and reshapes the role *son-man*, as only one of the various sons of universe-earth reality. Egoistic and hiper-individualistic ethical approaches shall be substituted by more equilibrated ways of balancing individual and collective interests and dynamics.

6. Participatory processes shall allow the questioning the classic liberal and neo-liberal economic approach, the promotion of more ecological and solidary types of economy and economic options, the discussion and consideration of other concepts and kinds of markets as well as other possible economic systems. The foundations and structures of world economic order (MB and IMF) as well as the control of bank activity shall be extensively discussed in order of being modified to avoid speculation and new economic crisis like the last one.

7. Participation shall be oriented by a new kind of relationship with environment and nature based in an attitude of humility and equilibrium. Some tendencies like very concentrated littoral urbanization, as well as extensive and inadequate soil use, traditional carbon/petroleum dependent energetic politics, classical tendency for over-exploitation of resources and careless deposition of residues, etc., have to be overcome. New scientific and technological approaches will help very much in this direction.

References

Agrawal A, Gibson C (1999) Enchantment and disenchantment: the role of community in natural resource conservation. World Dev 27(4):629–649. http://www.dgroups.org/file2.axd/.../agrawal_communities_in_NRM.pdf. Accessed 22 Nov 2010

Amorim D, Sato M (2010) O Processo de Participação nas políticas Públicas como Aprendizado de uma Ecologia de Resistência. sistemas.ufmt.br/ufmt/6b875539-412b-4bf5-9e74-9932580da917.docx. Accessed 22 Sept 2010

Aref A (2010) Community participation for educational planning and development. Nat Sci
    8(9):1–4. http://www.sciencepub.net/nature/ns0809/01_2335_ns0809_1_4.pdf. Accessed 10
    Nov 2010
Arnstein S (1969) A ladder of citizen participation. J Am Plan Assoc 35(4):216–224. http://lith-
    gow-schmidt.dk/sherry-arnstein/ladder-of-citizen-participation.html. Accessed 11 Nov 2010
Bavaresco A (2009) O pjê e a Cartografia: Os mapeamentos participativos como ferramenta ped-
    agógica no diálogo entre saberes ambientais. Dissertação de Mestrado. Brasília. http://www.
    trabalhoindigenista.org.br/.../BavarescoA_O-Pje-e-a-Cartografia_mapeamentos-participati-
    vos.pdf. Accessed 12 Nov 2010
Blackmore C (2006) In: Langevel H, Rolings N (eds) Where do participatory approaches meet
    social learning systems in the context of environmental decision making? http://oro.open.
    ac.uk/3294/. Accessed 13 Nov 2010
Blamey R, James R, Smith R Niemeyer S (2000) Citizens' juries and environmental value assess-
    ment. http://cjp.anu.edu.au/docs/CJ1.pdf. Accessed 10 Nov 2010
Bowen G (2008) An analysis of citizen participation in anti-poverty programmes. Commun Dev
    J 43(1):65–78. http://www.thecyberhood.net/documents/papers/poverty.pdf. Accessed 13 Nov
    2010
Buttoud G (1999a) Negotiation methods to support participatory forestry planning. In: Niskanen
    A, Väyrinen J (eds) Regional forest programmes: a participatory approach to support forest
    based regional development. EFI Proc 32:29–46
Buttoud G (1999b) Principles of participatory process in public decision making. In: Niskanen A,
    Väyrinen J (eds) Regional forest programmes: a participatory approach to support forest based
    regional development. EFI Proc 32:11–28
Campbell L, Vainio-Mattia A (2002) Participatory development and community-based conserva-
    tion: opportunities missed for lessons learned? Hum Ecol 31(3):417–437
Carr J (2006) Discourses of widening participation and social inclusion. PhD thesis. The Open
    University. http://oro.open.ac.uk/21326/1/Final.pdf. Accessed 13 Nov 2010
Collins K, Ison R (2006) Dare we jump off Arnstein's ladder? Social Learning as a new policy
    paradigm. In: Proceedings of PATH (Participatory Approaches in Science & Technology) Con-
    ference, 4–7 June 2006, Edinburgh. http://oro.open.ac.uk/8589/. Accessed 14 Nov 2010
Costa C, Pascoal M (2006) Participação e Políticas Públicas na Segurança Alimentar e Nutricional
    no Brasil. In: Albuquerque MC (ed) Participação popular em políticas públicas: espaço de
    construção da democracia brasileira. Instituto Pólis, São Paulo, pp 97–108. http://www.direito.
    caop.mp.pr.gov.br/arquivos/File/PoliticaHabitacionaInoBrasil.pdf. Accessed 10 Nov 2010
Dreyer L (2000) Disappointd of participation: finding the correct role for community participation.
    http://www.awiru.co.za/pdf/dryerlynette1.pdf. Accessed 15 Nov 2010
Enggrob B (1999) Five analytical frameworks for analysing public participation. In: Niskanen A,
    Väyrinen J (eds) Regional forest programmes: a participatory approach to support forest based
    regional development. EFI Proc 32:47–62
Finger-Stich A, Finger M (2003) State versus participation. Natural Resources Management in
    Europe. IIED & IDS, London
Freitas M (2008) Natureza, cultura, ambiente e desenvolvimento: um ensaio sobre a possibilidade
    sobre uma cultura (comum) da sustentabilidade. In: Parente T, Magalhães H (eds) Linguagens
    Plurais: cultura e meio ambiente. pp 13–40
Freitas M, Annunciato D, Nardi I (2009) A Mediação como Prática Discursiva Transformadora.
    Um estudo de caso com comunidades piscatórias ribeirinhas de Botucatu (SP, Brasil). In: Brito
    B, Alarcão N e Marques J (eds) Desenvolvimento Comunitário das Teorias às Práticas. Tur-
    ismo. Ambiente e Práticas Educativas em São Tomé e Príncipe. Gerpress Comunicação Empre-
    sarial e Marketing Lda., Lisboa, pp 192–204
Gonsalves J, Becker T, Braun A, Campilan D, De Chavez H, Fajber E, Kapiriri M, Rivaca-Cami-
    nade J, Vernooy R (eds) (2005) Participatory Research and Development for Sustainable Agri-
    culture and Natural Resource Management: A Sourcebook. Doing Participatory Research and
    Development 3. International Potato Center-Users' Perspectives With Agricultural Research

and Development, Laguna, Philippines and International Development Research Centre, Ottawa, Canada. http://www.idrc.ca/openebooks/183-3. Accessed 9 Nov 2010

Guiddens A (2005)

Hesselink F, Goldstein WE, van Kempen P, Garnett T, Dela J (2007) Communication, Education and Public Awareness (CEPA). A toolkit for the convention on biological convention. IUCN, Montreal. http://data.iucn.org/dbtw-wpd/edocs/2007-059.pdf. Accessed 9 July 2009

Hunsberger C, Kenyon W (2006) How to better represent different values in participatory approaches: a conference perspective. Proceedings of conference. Edinburgh, Scotland, 4–7 June 2006. http://www.macaulay.ac.uk/PATHconference/outputs/PATH_closing2.pdf. Accessed 13 Nov 2010

Innes J, Booher D (2004) Reframing public participation: strategies for the 21st century. Plan Theory Pract 5(4):419–436. http://escholarship.org/uc/item/4gr9b2v5#page-3. Accessed 15 Nov 2010

James R, Blamey R (1999) Public participation in environmental decision-making—Rhetoric to reality? Paper presented at the International Symposium on Society and Resource Management. Brisbane, Australia, 7–10 July. http://cjp.anu.edu.au/docs/appendix2.pdf. Accessed 19 Nov 2010

Kenyon W, Hunsberger C (2006) Involving people in policy development: a conference view. Proceedings of Conference. Edinburgh, Scotland, 4–7 June 2006. http://www.macaulay.ac.uk/PATHconference/outputs/PATH_closing1.pdf. Accessed 13 Nov 2010

Kienberger S, Steinbruch F (n d) P-GIS and disaster risk management: assessing vulnerability with P-GIS methods—Experiences from Búzi, Mozambique. http://www.uni-salzburg.at/pls/portal/docs/1/215961.PDF. Accessed 9 Nov 2010

Kouplevatskaya-Yunusova I (2005) The evolution of stakeholders' participation in a process of forest policy reform in Kyrgyz Republic. Schweiz Z Forstwes 156(10):385–394

Kouplevatskaya-Yunusova I (2006) The involvement of stakeholders in a forest policy reform process: democracy promotion and power redistribution. Schweiz Z Forstwes 157(10):483–490. http://www.atypon-link.com/SFS/doi/pdf/10.3188/szf.2006.0483. Accessed 9 Nov 2010

Kouplevatskaya-Yunusova I (2007) Participation as a new mode of governance?: scientists and policy makers in a double spiral. In: Reinolds K, Thomson A, Köhl M, Shannon M, Ray D, Rennolls K (eds) Sustainable forestry: from monitoring and modelling to knowledge management and policy science. CABI Internationalp, Cambridge, pp 35–55

Leite S, Freitas M, Seneca A (2005) Promoting education for sustainable development through communitarian problem solving: a case study in the national park of Peneda-Gerês. In: Leal Filho W (ed) Handbook of sustainability research. Environmental education, communication and sustainability. Peter Lang, Frankfurt a. M., pp 595–622 (Chap. 23)

Mannigel E (2002) Participatory solution of land use conflicts in protected area management in the Brazilian Atlantic Forest. http://www.tropentag.de/2002/abstracts/full/36.pdf. Accessed 14 Nov 2010

Maranhão T, Teixeira A (2006) Participação no Brasil: dilemas e desafios contemporâneos. In: Albuquerque MC (ed) Participação popular em políticas públicas: espaço de construção da democracia brasileira. Instituto Pólis, São Paulo, pp 109–119, 124. http://www.direito.caop.mp.pr.gov.br/arquivos/File/PoliticaHabitacionalnoBrasil.pdf. Accessed 12 Nov 2010

Maturana H, Varela F (2002) A árvore do conhecimento. As bases biológicas da compreensão humana. Editoria Palas Athenas, São Paulo

McCall M (2004) Can participatory-GIS strengthen local-level spatial planning? Suggestions for better practice, GISDECO 2004, Skudai, Johor, Malaysia, 10–12 May 2004. http://www.iapad.org/publications/ppgis/Mike_McCall_paper.pdf. Accessed 12 Nov 2010

Mészáros I (2010) Atualidade histórica da ofensiva socialista. Editorial Boitempo, São Paulo

Mohan G (2006) Beyond participation: strategies for deeper empowerment. In: Cooke B, Kothari U (eds) Participation: the new tyranny? Zed Books, London, pp 153–167. http://oro.open.ac.uk/4157/1/TYRANNY3.pdf. Accessed 13 Nov 2010

Paizano J, Jardinet S, Urquijo J (2006) Madrid (UPM). http://www.iapad.org/publications/ppgis/afc_06_sigp_las%20sabanas.pdf. Accessed 14 Nov 2010

Soros G (1999) A crise do capitalismo. As ameaças aos valores democráticos. As soluções para o capitalismo global. Editora Campus, Rio de Janeiro, pp 1–315

Soros G (2008) O novo paradigma para os mercados financeiros. A crise atual e o que ela signific. Editora Agir, Rio de Janeiro, pp 1–207

Toro J, Werneck N (2004) Mobilizaçao Social: um modelo de construir a Democracia. Editora Autêntica, Rio de Janeiro

Tritter J, McCallum A (2006) The snakes and ladders of user involvement: moving beyond Arnstein. Health Policy 76:156–168. http://www.engage.hscni.net/library/The%20Snakes%20 and%20Ladders%20of%20User%20Involvement.pdf

Wakeford T, Singh J (2008) Towards empowered participation: stories and reflections. London. Particip Learn Action 58:6–9

Warburton D (1997) Participatory action in the countryside. A literature review for the Countryside Commission, August 1997. http://www.sharedpractice.org.uk/Downloads/Participatory_Action_Review.pdf. Accessed 25 Nov 2010

Weiner D, Harris T, Craig W (2002) Community participation and geographic information systems. Paper presented in the Conference on International Agricultural Research for Development, Deutscher Tropentag, Witzenhausen, 9–11 Oct 2002. http://oook.info/amazon/WeinerEtAl.pdf or http://ebookbrowse.com/community-participation-and-geographic-information-systems-pdf-d1292704. Accessed 13 Nov 2010

Uemura M (1999) Community Participation in Education: What do we know. Prepared by Mitsue for Effective Schools and Teachers and the Knowledge Management System HDNED, The World Bank. http://siteresources.worldbank.org/INTISPMA/Resources/383704-1153333441931/14064_Community_Participation_in_Education.pdf. Accessed 13 Nov 2010

Veja P, Freitas M, Álvarez P, Fleuri R (2009) Educación Ambiental e Intercultural para la sostenibilidad: fundamentos y praxis. Utopía y Praxis Latinoamericana. Año 14(44):25–38

Chapter 11
Dragonfly

Mateusz Banski

Abstract We know that, despite many efforts to save it, our environment is still deteriorating at an alarming rate. While many organisations and individuals want to reverse this trend by raising awareness the importance of preserving healthy ecosystems, very few people seem to know what concrete actions they can do. In the next few pages, I try to list some ideas, but more importantly, I invite you, the reader, to take responsibility, to take matters into your own hands, and to make the world around you a better place. Every individual action adds up, so question yourself: do I care? Inform yourself: what's my impact on ecosystems and people? And most importantly: act. There are good and bad ways to raise awareness, there are ways to "sell" biodiversity to those around you, to make it appealing, and there is a way to explain it to those who only swear by number$. It's about time that we start living in harmony with nature.

11.1 Introduction

On the 3rd of May 2010, I arrived in Braga to take part in the 2nd International School Congress: Natural Resources, Sustainability and Humanity, organised by the Lamaçães School Cluster in partnership with the Gualtar School Cluster and with the collaboration of the University of Minho. I was in Portugal to represent the United Nations Secretariat of the Convention on Biological Diversity (www.cbd. int). Please note however that the views expressed here are not necessarily those of the CBD, as I was asked to give my own[1] perspective of the Congress.

Speakers and participants seem to agree that, although education about biodiversity and raising awareness about the consequences of its loss are essential means to reach the general public, it's time to do concrete actions. Yet, nobody seems to know where to start.

---

[1] "Dragonfly" became my nickname during the Congress. Best wishes to my 'insect' family!

M. Banski (✉)
Secretariat of the Convention on Biological Diversity, 413, Saint Jacques Street, suite 800, Montreal QC H2Y 1N9, Canada
e-mail: mateusz.banski@cbd.int

A. Mendonca et al. (eds.), *Natural Resources, Sustainability and Humanity,*   167
DOI 10.1007/978-94-007-1321-5_11, © Springer Science+Business Media Dordrecht 2012

Fig. 11.1 In 2010, the International Year of Biodiversity, I was very privileged to meet so many bright and committed minds gathering in Braga. The 2nd International School Congress brought together specialists from a range of thematic areas putting a biodiversity spotlight on very interesting topics such as human psychology and behaviour, philosophy and education, natural disasters, water scarcity and management, to name only a few.

We know that, despite many efforts to save it, our biodiversity and environment are still deteriorating at an alarming rate, especially freshwater wetlands, sea ice habitats, salt marshes, coral reefs, seagrass beds and shellfish reefs[2].

In fact, for the first time, I start to wonder, perhaps even worry, about my own future; not about my career, material needs or retirement plans, but about basic wellbeing. Taking it a step further: do I want to bring a child into such a world? Putting it at risk of respiratory problems, allergies, accumulation of various chemicals in its tiny body? Knowing that by the same I will contribute to the growth of world population, adding more pressure on our overexploited planet and ecosystems? The coming 20–50 years will be very interesting indeed: a turning point or a point of no return? (Fig. 11.1).

On one hand, some of us live in an industrialized world, where people have become accustomed (addicted?) to commodities, to "easy living" without self-awareness, and consuming huge amounts of natural resources (energy, water, habitat, etc.). In fact, more than 50% of human population now lives in urban areas. Many of such "daily needs" have a tremendous ecological footprint, clearly unsustainable. A coffee here, a soda there, the plastic cup goes in the trash. Multiply that by an entire city… little things add up. Of course some of us (a minority of the population) are already more conscious of our impact and make a true effort to reduce our

---

[2] For most accurate and recent data, please refer to the Global Biodiversity Outlook 3, by the Convention on Biological Diversity (gbo3.cbd.int).

ecological footprint, keeping in mind that modern society imposes certain standards and lifestyle which still need to change.

On the other hand, the majority of the world's population strives to achieve the same high living standards… it would only be fair for them to live *la dolce vita* as well. We all know how the story goes.

But here I am just talking again. What can I DO?

We look at others, at the government(s), at our society, our neighbours, but do we really look at ourselves? Or do we say: "poor pandas, that's horrible" and turn the TV off to go on with our lives?

Step 1—Question Yourself. Do I Care? Do I care about the environment? Most of us do, at least in theory. Do I care about myself? Do I want to be happy? That's pretty easy. Do I care about my family and friends? Do I want them to be healthy and have a good life? Yes, of course. Do I care about other people? Do they also deserve to live healthy and happy lives? Do I care about the poor farmer, sick from all the pesticides he inhaled to produce food or fibre for me? Do I really? This is where it gets more difficult, but I should care. I'm sure that it's not his dream job: he's breathing in this poison so that his children can have a fraction of the life that I have.

I think that it's all about balance, altruism, compassion and solidarity; words we seldom hear in 'mainstream' society unless there's a cataclysm of some sort. We all protect what is ours; our health, our family, our home, our country… why not our planet? Just because a piece of land is not our property or isn't within our political borders, it's not our problem anymore?

Take any Hollywood movie about an alien invasion of the Earth: human beings unite to save their lives and their resources. In that case citizenship, religion, skin colour and beliefs don't seem to matter—it's OUR planet and we all fight for it. Why does it work in the movies?

Step 2—Inform Yourself. What's My Impact on Ecosystems and People? In today's world, there is no excuse for remaining uninformed or misinformed.

Everything we do has direct and/or indirect repercussions on the environment near and far, which in turn has effects on humans that inhabit these environments, including an impact on ourselves.

Human beings, as a species, rule the Earth. We have great power to alter our environment and in doing this we can greatly impair ecosystem functioning and put other species at risk. But with this power comes the responsibility to care for them and to protect them… remember: we are the good guys in the aliens' movies.

Homo sapiens ("wise man", I sometimes wonder…) evolved in nature for hundreds of thousands of years. Now, in a matter of decades, we are so drastically altering the systems that make our life possible that we are putting ourselves at risk. Remember Easter Island! Ecosystems require a balance to function properly and to keep providing us with countless services[3] (Fig. 11.2).

---

[3] Ecosystem services such as food and pollination, fuel, medicine, building materials, clean air and water, soil formation, protection from natural disasters, culture & spirituality, and many, many more. See the CBD's factsheets for more details: www.cbd.int/2011-2020/learn/factsheets.shtml

Fig. 11.2 Braga gives me a wonderful sensation of lightness. I feel at home. Santa Bárbara is a beautiful garden, densely planted with colourful and sweet smelling flowers. In the background, at the far end, you can see the Paço Medieval de Braga (fourteenth and fifteenth centuries) and its stone arches. Breathtaking! Biodiversity is beautiful and amazes us all, and it gives cultural, artistic and spiritual inspiration—an example of ecosystem services

Today, this balance needs to go beyond ecosystems because more and more humans are in the equation. A person, with barely enough to eat, has one priority: to survive for one more day, and they will likely do whatever it takes, even if it means eating the last remaining plant or animal. On the other extreme, there's the person who never has enough. Everyone else falls somewhere in between, and only very few stand in the middle, where we should all be.

More than ever, we are connected to all other human beings on the planet. For example, most of us depend on farmers who provide our food. A person made the chair that you are sitting on. Therefore we should ensure their health and wellbeing. For example, climate change and desertification could affect that same farmer: not only he will loose the proper conditions to produce our food, but if his environment becomes too harsh, he may be forced to move to another area. He would then become an environmental refugee and may ask for refuge in your country—adding pressure to a system that most probably has already reached its limit. Or just think of all the conflicts that start over resources. We ALL depend on our planet's finite natural resources and we must learn to share them.

fixed amount of resources / growing population = less resources per person

Some say that science will solve it all. Then why do they not listen to the same science that warns us of the dramatic consequences of overexploitation and destruction of habitats?

Am I willing to have less and less over time? Do I still care about others? Do I still care about the future generations? I think I do, and it's time to do something about it. The critical problem is that it is easy to reach this point of reasoning, but very difficult to overcome the feeling of hopelessness and impotency that immediately follows it. Well, in the next section I'll share some of my first steps forward and challenge you to do the same.

Step 3—Act. What Can I Do? First, I stop waiting for the government(s), for society and others to do something. What can I do myself, starting right now?

I hear people around me say: "I compensate for this or that because I recycle". That's not good enough.

Belgium published *366 actions for biodiversity*, and its abridged version *52 actions for biodiversity*[4] (www.jedonnevieamaplanete.be). This is an excellent starting point: daily tips to help the environment. I will base myself partly on that to change my behaviour. Who will join me in doing the same?

1. I will progressively make a list of my food and I will research its origin, ingredients and production methods. I will buy food only if it isn't an endangered species (ref. www.iucnredlist.org, www.cites.org, www.msc.org) and if its production doesn't have a negative impact on the environment. I'll eat local, sustainable and organic as much as possible. I won't buy exotic and/or invasive species, which are produced to the detriment of native fauna and flora. Different countries have different organic certification schemes, and it's important to differentiate between "100% organic", "less-than-100% organic" and "containing organic ingredients" because many companies use the term for marketing. I will also apply this to seafood, forest products or ingredients that have different or no certification mechanisms. For example, I will avoid palm oil of uncertain origin: tropical forests are being replaced by palm trees to produce oil. Some of these forests are home to the last orang-utans.

2. I'll minimise consumption, energy consumption (air-conditioning, heating, etc.) and waste (reduce, reuse, recycle, compost, etc.). I'll avoid over-packaged products and items that are damaging to the environment either through their components (metals, plastics, chemicals, etc.), production method, or which are harmful to the people who made them (health, working conditions, etc.). In other words, I'll buy natural, certified, reused/recycled products as much as possible. For example, I will make a big effort to save and recuperate water (dishes, gardening, washing, etc.). This could partly consist of recuperating rain water for the garden or, when I buy a house, using rainwater for the toilet. But most of all, I'll buy only what I really need. Do I really need the latest mobile phone or computer model? Coltan is a mineral used in electronic equipment. It's extracted mainly in the Democratic Republic of Congo. Very often these mines are damaging to the regional biodiversity: illegal hunting of endangered species (gorilla, elephant, okapi, etc.) to feed the workers; massive deforesta-

---

[4] The booklet *52 actions for biodiversity* has been translated into the six UN languages and Japanese, and is available in all 27 languages of the European Union.

tion to provide fire wood and construction materials; erosion, pollution and armed conflicts. The longer we keep our old electronics, and the more we recycle them and their components, the less we contribute to these problems.

3. I will use natural, non-toxic, recycled and durable supplies and furniture. I will buy FSC and PEFC[5] certified products only. I will be careful when I print, to use both sides of the sheet, not to waste paper or ink (or anything else for that matter), and I will not print anything unnecessarily. I will bring my lunch in reusable containers, and my own utensils and cup. I'll turn off my computer at the end of the day. All these small actions add up.

The Programme for the Endorsement of Forest Certification (PEFC), the world's largest forest certification system, ensures that a product comes from environmentally and socially responsible and economically viable forestry. The label itself however is not as easy to find because many countries use their own names, for example: Sustainable Forestry Initiative (SFI), American Tree Farm System (ATFS), Canadian Standards Association (CSA), Australian Forestry Standard (AFS), Malaysian Timber Certification Council (MTCC), etc.

PEFC is the only international forest certifier based on different international conventions and it requires adherence to all eight core ILO conventions, in addition to other international conventions relevant to forest management (ex. Convention on Biological Diversity).

The Forest Stewardship Council (FSC) is an international NGO that sets the standards and promotes responsible forest management. It's the main alternative forest certification system. The FSC logo[6] therefore ensures that a product is sustainable. The label can be found on a wide variety of timber and non-timber products such as paper, furniture, medicine and jewellery.

The two systems don't recognise one another, even though they use the same forest management standard in some countries (ex. UK, Switzerland and Norway).

4. I will use non-polluting household products and cosmetics. Many detergents and body care products (soap, sun block, etc.) contain chemical products (phosphates, solvents, perfumes, preservatives, etc.), which often end up in rivers even when the waste water is treated. They cause pollution and affect fauna and flora. Eco-friendly and natural cleaning products are easy to find and to make (Marseilles soap, water and vinegar, natural extracts, even toothpaste, etc.).

5. Around the house, the garden and other property, even if it's only a balcony or a window, I'll create habitat, nesting sites and green corridors to integrate endemic species as much as possible. I'll create habitat to encourage colonisation by invertebrates such as bees, wasps[7], ants, butterflies, ladybugs, spiders, earwigs and worms, which pollinate flowers, help maintain the soil and protect

---

[5] FSC and PEFC are the main two forest certification systems.

[6] Some accuse the label of greenwashing, but I think that when an organization becomes this big it's difficult to control everything and everyone.

[7] Can you tell the difference between a bee and a wasp? You'd be surprised how many people can't!

the garden from pests. I'll use eco-friendly and certified building materials, which are durable and recyclable. Even paint can be natural and non-toxic. I'll try to follow green architecture principles, using the sun and the environment as sources of clean energy.

6. Whatever I can't reuse or recycle, I will dispose of in a safe manner. Electronic equipment, batteries, cigarette butts, plastic, and many other products (or their components) are toxic to animals and plants; they can infiltrate the soil and contaminate ground water. These compounds eventually resurface and end up on my plate and in my drinking water. Such a vicious cycle causes additional negative effects. For example, amphibians face the greatest risk of extinction (water pollution, habitat destruction, desertification and many reasons). Where the frogs have disappeared, people now have to use pesticides to control disease-carrying insects: these chemicals affect ecosystems and are toxic to local human populations, which adds to their problems.

7. I will continue to use my BMW (Bus-Metro-Walk), bicycle, public transport and carpooling. I'll avoid taking the airplane when I have other alternatives, and when I can't I'll compensate for my CO_2 emissions by supporting reforestation projects. I will support sustainable tourism and ecotourism. For example, I'll support the local people by eating their food, but I'll refuse to eat dishes made from endangered species (ex. turtle, red tuna, sea cucumber, bushmeat, etc.) or when the harvesting method is unsustainable or damaging the environment. The same goes for traditional medicine, souvenirs and other products (ex. rhinoceros, elephant, tiger, musk deer, etc.).

   Sustainable tourism applies to all forms of tourism, not only to natural areas, and it aims to have a positive impact on the local people, environment and culture.

   Ecotourism is "responsible travel to natural areas that conserves the environment and improves the well-being of local people" (The International Ecotourism Society, www.ecotourism.org). It follows certain principles, such as: minimal impact, respect of the environment and culture, bringing financial benefits to the environment and to the local people, promotion of local society and politics.

   Many tour operators use the term "ecotourism" wrongfully. To date, out of about 125 sustainable tourism certification programmes, there are perhaps only 15 specific for ecotourism, and only a handful of countries have national ecotourism certification programmes: Australia, Botswana, Estonia, Ireland, Kenya, Mexico, Namibia, Norway, Romania, Sweden and the EU (www.rara-avis.com).

8. When I was a child, I loved nature so much that I wanted to bring it home. Now I know that it's the worst thing to do, first because I'm removing the species from its natural habitat and second, because I'm potentially bringing it into a new habitat where it could become invasive. I will tell those around me to leave the species where they belong, especially when dealing with endangered or island species. I will only get souvenirs, which are not made from endan-

gered species, or which are produced in a sustainable way (coral, ivory, turtle, Brazilian rosewood, etc.), and which have a CITES permit.

9. I will continue to raise awareness around me. If I wasn't already working for an environmental organisation, I would volunteer for one or support it in another way. Through communication tools such as internet, blogs, letters, TV and radio, I will also support and encourage politicians and other public figures that are doing concrete actions for the environment. For example, ambitious policies such as the creation of protected areas, management of invasive species, certification mechanisms, conservation projects, etc.

10. I will redouble efforts in marketing (see below) and mainstreaming biodiversity and ecosystem services.

Simply put, ecosystems supply us with resources (raw materials) and do countless other things for us, directly and indirectly, so much discreetly that we often forget that they are there. Ecosystems moderate weather extremes and their impacts (ex. drought, floods, etc.); mitigate climate change; absorb and store CO_2; protect water channels and shores from erosion; regulate disease-carrying organisms; provide ingredients for pharmaceutical, biochemical and industrial products; are a source of energy and biomass fuels; decompose waste and detoxify pollution; generate, maintain and renew soil fertility (nutrient cycling); pollinate crops and plants, and disperse seeds; control agricultural pests and diseases; produce food (crops, wild foods and spices, seafood, etc.); produce wood and fibre; produce oxygen, purify air and water; give cultural, intellectual, artistic and spiritual inspiration; allow recreation (ex. ecotourism); hold the answers to scientific questions; hold the cures to diseases; and many more.

For example, take a fruit. The fruit develops from a flower, which was pollinated by a bee, other insect or bird. It took a plant to produce the flower. The plant needs water and soil. Soil is largely made from organic materials (leaves, dead organisms, etc.) and maintained by decomposers (fungi, worms and other invertebrates). The pollinators and decomposers have their own requirements (food, water and shelter). If any part of this chain is missing or damaged, the other elements are affected as well, and the end result (in this case your food) is put in jeopardy.

11. I will be an example for others. Each of us is in a good position to tell and influence others (ex. family, friends, students, community, population, fans, etc.). If you are a teacher or a public figure, you are in a privileged position to change the way people think and act. If you are a parent, be a good example for your children, it's in their interest above all. If you are a child, show your parents that you want a clean environment.

Of course, this list is not exhaustive. I will make more changes after I get accustomed to my new lifestyle, for example: reduce my meat consumption (which is already low). Evolution takes time.

11.2   Marketing Biodiversity

Today's mainstream society seems to run mainly on money and glamour. Therefore, going back to what Joan Freeman had said during her presentation, it's time that we make nature fashionable—the hip thing to do! Marketing biodiversity should be a mixture of information, feelings and beliefs. Our goal is to change behaviours and to redefine what is socially acceptable so that **unsustainable** becomes **unacceptable**.

> Mr. Businessman, your right to make profit stops where it infringes on my right to clean air, water and food.

11.3   The Economic Argument

The Economics of Ecosystems and Biodiversity (TEEB) is a project that uses economics as an argument for conservation of ecosystems and biodiversity. This scientific study shows that our current economic model is outdated by some 200 years, when human population was relatively small and natural resources seemed unlimited. In fact GDP doesn't reflect "many vital aspects of national wealth and wellbeing, such as changes in the quality of health, the extent of education, and changes in the quality and quantity of our natural resources"[8].

> Not all that is very useful commands high value (ex. water) and not everything that has a high value is very useful (ex. diamond). (Adam Smith 1776)

Much too often, ecosystem services are taken for granted and their monetary value is not calculated, and therefore they are not taken into consideration by conventional development projects. The most common result is their degradation, with consequences for those who rely on them.

Although ecosystem services are still a new concept, and not fully understood or measured, TEEB puts monetary values on some of these services, giving concrete examples of countries and policies where conserving biodiversity (prevention) is much more profitable than repairing the damages or trying to replace such services.

Economics are certainly a strong argument for decision-makers, CEOs and everyone else who swears by number$. Still, it's not easy to "sell" biodiversity to them. In parallel, we should try to appeal to the masses through the positive feelings that nature gives us all.

> One touch of nature makes the whole world kin. (William Shakespeare 1602)

It seems that both Adam and Bill were onto something back then. A study on biodiversity branding by Futerra Sustainability Communications shows that bringing out the positive emotions in people gives much better results than pounding them with negative messages. "Consumer brands don't just sell products, they sell a set of brand values and promises which resonate powerfully with specific people. And they are incredibly good at it. Biodiversity deserves the same success"[9].

---

[8] The Economics of Ecosystems and Biodiversity (TEEB) www.teebweb.org
[9] Futerra.

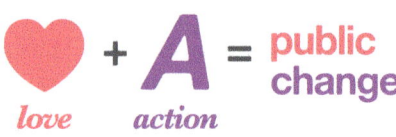

Fig. 11.3 *Less loss:* Kill the extinction message. Loss generates apathy, not action *More love:* Celebrate our love of nature. It is the most powerful driver of public behaviour. *Target need:* Use the need message wisely. It's often not right for public consumption, but it's the cornerstone of policy and business decisions. *Add action:* Always partner love and need messages with Action. Once your audience is inspired, they will want to know what to do.

Futerra has published a booklet[10] "for biodiversity campaigners, policy makers and media who are open to radically changing our biodiversity message, in order to radically increase action". They sum it up in the figure above (Fig. 11.3):

It really does work: I experienced something similar on a small scale. I posted two videos on facebook: the first, very artistic and well made, but also very sad, showing that we are losing biodiversity; the second an amateur clip of seahorses "dancing" to the sound of music. Nobody commented on the first one, while a dozen people liked the seahorses and were all in awe. I believe that this is the way to go.

Even though I grew up loving nature, I studied ecology and I work to help our biodiversity, I still have much to learn. Moreover, when I speak with people, I realise that many of them are simply not aware. The next 20–50 years will be very interesting ones as they will determine whether society adopts new values or if we do hit that point of no return.

I thank you for your attention. If you are already part of us—people who truly respect the environment and, by the same, all human beings—it's time to reach those beyond our usual circles. And if any of what you read here today is new to you, but makes sense, welcome! If it doesn't make sense, please let me know why. I truly hope that by our actions, by becoming models for others and by bringing out their positive feelings about the environment, a new way of thinking will emerge; a new "mainstream society".

---

[10] www.futerra.co.uk/downloads/Branding_Biodiversity.pdf

2011–2020 is the United Nations Decade on Biodiversity, starting with the International Year of Forests in 2011, let this be the beginning of a new life. We have 10 years to change mentalities and behaviours, our own and those of others: www.cbd.int/2011-2020.

11.4 Thank You

A big thank you to Portugal and to the Portuguese Committee for Biodiversity for their excellent participation in celebrating the International Year of Biodiversity!

E muito obrigado a Angela Mendonca, Maria José Marques, Nuno, Marta and all the good people of the Lamaçães and Gualtar School Clusters for a super congress, great organisation and, above all, for their kindness, hospitality and hard work to make our world better. I send my best wishes to all of you!

Mateusz

The "Heart that loves all people" made by 5-year old children of the Meadala Kindergarden, Viana do Castelo, and their teacher Maria da Conceição Carvalho. It was made into a silver jewel by Dr. Freitas.

11.5   The Convention on Biological Diversity

Opened for signature at the Earth Summit in Rio de Janeiro in 1992, and entered into force in December 1993, the Convention on Biological Diversity (www.cbd.int) is an international legally-binding treaty with three main goals: conservation of biodiversity; sustainable use of biodiversity; fair and equitable sharing of the benefits arising from the use of genetic resources. Its overall objective is to encourage actions which will lead to a sustainable future.

The conservation of biodiversity is a common concern of humankind. The Convention on Biological Diversity covers biodiversity at all levels: ecosystems, species and genetic resources. It seeks to address all threats to biodiversity and ecosystem services, including threats from climate change, through scientific assessments, the development of tools, incentives and processes, the transfer of technologies and good practices and the full and active involvement of relevant stakeholders including indigenous and local communities, youth, NGOs, women and the business community. The Cartagena Protocol on Biosafety, a supplementary treaty to the Convention, seeks to protect biological diversity from the potential risks posed by living modified organisms resulting from modern biotechnology.

The CBD's governing body is the Conference of the Parties (COP). This ultimate authority of all governments (or Parties) that have ratified the treaty (currently 193) meets every 2 years to review progress, set priorities and commit to work plans.

The Secretariat of the Convention on Biological Diversity (SCBD) is based in Montreal, Canada. Its main function is to assist governments in the implementation of the CBD and its programmes of work, to organize meetings, draft documents, coordinate with other international organizations and collect and spread information. The Executive Secretary is the head of the Secretariat.

11.6   Message from the Executive Secretary

The work of the 2nd International School Congress is of particular interest to the Convention on Biological Diversity as it encompasses one of the key aspects of the Communication, Education, and Public Awareness (CEPA) programme of work, which is to educate people on the importance of conservation and sustainable development, as well as the dangers associated with neglecting this duty. Although the Secretariat and the 2nd ISC play very different yet equally important roles in promoting the preservation of the Earth's precious resources, I believe we can work together to achieve our shared goals. Having a CBD representative attend this unique meeting builds synergies and brings us all one step closer to fulfilling our objectives, thus contributing to a better and more sustainable future for all.

The Secretariat of the Convention on Biological Diversity is very grateful to Portugal, to the 2nd International School Congress and to Angela Mendonca for their major contribution to the International Year of Biodiversity. I also wish to express our utmost gratitude to Angela, to the children of the Meadela Kindergarten in Viana do Castelo and their teacher Maria da Conceição Carvalho, the craftsmen, the scientists and researchers, the teachers and students, and all the institutions involved in the 2nd ISC for sending Portugal's first gift to the CBD Museum of Nature and Culture: an exquisite silver pendant called the "Heart that loves all people". Viana's heart now shines at the CBD Secretariat in Montreal and is admired by visitors from around the world.

The CBD Museum of Nature and Culture was created in early 2006 for the purpose of raising awareness of the intimate relationship between humanity and the environment, in which humans live through various forms of artistic expression reflecting the unique cultural and biological diversity of the various parts of our world. Your donation is undoubtedly a true testimony to your country's commitment to the conservation of biological diversity.

The International Art Competition "When the world becomes a Canvas" was yet another creative initiative of the 2nd International School Congress, coordinated by the Portuguese UNESCO National Commission and in partnership with the Institute for the Conservation of Nature and Biodiversity. It reached far beyond Portugal's borders: students of all ages and nationalities participated in the competition.

This contest was launched by the Lamaçães Schools Cluster to raise public awareness and to encourage communities to protect the environment. Simply put, the contest promotes the use of traditional techniques, non-toxic and natural inks and other environmentally-friendly media as an artistic support... leading to amazing results.

The 1st prize winners of the two categories, students from Mozambique and Brazil, were published in National Geographic Magazine and on the International Year of Biodiversity website. The International Art Contest is also part of the Earth Day Celebrations under the theme of the International Year of Biodiversity.

Finally, I also wish to congratulate the 2nd ISC for organizing the fundraiser operatic recital performed by soprano Stella Mendonca to support the project: 'Building a School, building a Future', to build a school in the Gorongosa National Park, in Mozambique. The story of the Gorongosa National Park is touching and it is an example for everyone to follow on how to transform a devastated land into a healthy and flourishing ecosystem.

I smile when I am reminded of the title of the opening song: "Alma grande e nobil core" (Great soul and noble heart): it perfectly describes Angela. I applaud your outstanding work.

Ahmed Djoghlaf
Executive Secretary of the Convention on Biological Diversity (2006–2012)

References

Futerra Sustainability Communications (2010) Branding biodiversity, the new nature message. Futerra Sustainability Communications, London. www.futerra.co.uk

Rara Avis Rainforest Lodge & Reserve (2010). www.rara-avis.com

Royal Belgian Institute of Natural Sciences, Degueldre C (2010) 52 Actions for Biodiversity. Royal Belgian Institute of Natural Sciences, Brussel. www.sciencesnaturelles.be/active/biodiv2010/biodiv2010_site/366tips

Royal Belgian Institute of Natural Sciences, Degueldre C (2010) 366 gestes pour la biodiversité. Royal Belgian Institute of Natural Sciences, Brussels. www.sciencesnaturelles.be/active/biodiv2010/biodiv2010_site/366tips

Secretariat of the Convention on Biological Diversity (2010) Global Biodiversity Outlook 3 is an open access publication, subject to the terms of the Creative Commons Attribution License. SCBD, Montreal. http://gbo3.cbd.int (ISBN-92-9225-220-8)

Secretariat of the Convention on Biological Diversity (2010) The Convention on Biological Diversity factsheets. SCBD, Montreal. www.cbd.int/2011-2012/learn/factsheets.shtml

Sukhdev P. et al (TEEB) (2008) The Economics of Ecosystems and Biodiversity: An Interim Report. European Commission, Brussels. www.teebweb.org

The International Ecotourism Society (2010) www.ecotourism.org

Chapter 12
Integrating Environmental Concerns into Development Assistance Policy

A Few Reflections

Alexandra Ferreira de Carvalho

> *Imagination is the beginning of creation. You imagine what you desire; you will what you imagine; and at last you create what you will.*
> (Bernard Shaw)

Abstract Today, it is inconceivable that environmental policy and spatial planning could exist without interaction with other social sectors. In order to achieve their goals, therefore, ministries with responsibilities in these matters, bearing in mind the cross-sectoral nature of their powers and from a perspective of policy cooperation and consistency, must take these sectors into account, specifically those relating to the economy, industry, agriculture, energy, transport, public works, health, education and tourism. In light of what we know today, incorporating environmental concerns into plans, projects, programmes and national strategies developed for these sectors, is of strategic importance for sustainable development. What I aim to do in the following lines, therefore, is to encourage joint reflection on the integration of environmental concerns into development aid policy and, more specifically, to discuss the importance of reinforcing environmental integration into the various development assistance policies.

My starting point for this essay was an article written by Dr. Luís Chainho and myself for a speech to be presented by the Office of International Affairs (GRI) at the 1st Conference of the Impact Assessment Portuguese Language Network on the theme of "Transport, Urban Development and Impact Assessment" in Lisbon in June 2010.

Today, it is inconceivable that environmental policy and spatial planning could exist without interaction with other social sectors. In order to achieve their goals, therefore, ministries with responsibilities in these matters, bearing in mind the

A. Ferreira de Carvalho (✉)
Office of International Affairs—Department of Foresight
and Planning and International Affairs of the Ministry for Agriculture,
Sea, Environment and Spatial Planning, 1200-433 Lisbon, Portugal
e-mail: alexandra.carvalho@dpp.pt

A. Mendonca et al. (eds.), *Natural Resources, Sustainability and Humanity,* 181
DOI 10.1007/978-94-007-1321-5_12, © Springer Science+Business Media Dordrecht 2012

cross-sectoral nature of their powers and from a perspective of policy cooperation and consistency, must take these sectors into account, specifically those relating to the economy, industry, agriculture, energy, transport, public works, health, education and tourism.

In light of what we know today, incorporating environmental concerns into plans, projects, programmes and national strategies developed for these sectors, is of strategic importance for sustainable development.

What I aim to do in the following lines, therefore, is to encourage joint reflection on the integration of environmental concerns into development aid policy and, more specifically, to discuss the importance of reinforcing environmental integration into the various development assistance policies.

The process that has brought us to where we are today has been a long one. Much of the abuse which our planet has been subjected to, and which, unfortunately, it continues to be subjected to, can be explained by the fact that we traditionally view natural resources as inexhaustible and capable of regenerating themselves at the same speed at which we consume and produce them.

Poverty reduction and environmental conservation are two of the main challenges facing mankind today and they are crucial to our attempts to achieve sustainable development.

As a general rule, poor countries exhibit high levels of environmental degradation. The lack of money to invest in solid waste treatment and sewage systems and in cleaner technologies results in the contamination of rivers, soil and air. Moreover, environmental problems, such as climate change, are responsible for increasing poverty by placing at risk many of the activities, which the poorest communities depend on, such as agriculture and fishing.

Climate change is one of our greatest challenges. A clear international consensus is forming regarding the likely impacts at the economic, social and environmental levels and the urgent need for action. This issue becomes even more important when we recognise that some of the nations that stand to lose most from climate change and its related impacts are some of the poorest developing countries, including the small island states. It is also these countries which are most vulnerable due to the combination of multiple exacerbating factors and a limited capacity to adapt. The effects of climate change are a major threat to the achievement of the Millennium Development Goals and a serious constraint on the effectiveness of actions and measures designed to fight poverty in the national strategies of the countries mentioned above.

As environmental problems are global problems, it is undeniable that the basis on which decisions are taken at the global level is affected not only by the economic and financial crises, but also by global challenges such as climate change, migration and threats to security.

But it has not always been this way. Less than 50 years ago, the issue of environmental protection was often seen as an obstacle to development, a drawback or constraint. "Environment" and "development" as concepts were commonly regarded as existing in mutually opposite camps.

As a result of the various warning signs and environmental disasters that the world has witnessed, this view has changed and environmental protection has become central to the international political agenda. And environmental policy has become regarded as a demanding aspect of research into nature and the development of new products.

The major international summits on the environment and development that took place in the 1970s, such as the Stockholm Conference on the Human Environment in 1972, clearly established the link between poverty and environmental degradation. The other major conferences that followed, the UN Conference on Environment and Development (1982), the Millennium Summit (2000), the World Summit on Sustainable Development (2002) and the Monterrey Conference on Financing for Development (2002), to name but a few, have forced us to constantly re-think new solutions and to strengthen consensus of the view that the environment and development are two sides of the same coin. And it is from the interaction of these that we will have to find solutions for the challenges facing mankind.

Integrating environmental concerns into development policy assumes the following:

- Discovery of new energy solutions, namely within the field of renewables, and greater energy efficiency;
- Sustainable use of natural resources;
- Development of climate change adaptation and mitigation solutions;
- Transfer of eco-efficient technologies;
- Fostering of informed debate and public participation to formulate policies on the impact of economic development on countries' environmental heritage.

The importance of integrating environmental considerations is essentially because economic and social development and the environment are fundamentally interdependent. The way in which we manage the economy and political and social institutions has a critical impact on the environment, in the same way that environmental quality and sustainability are vital for the performance of the economy and our social well-being. Environmental integration, therefore, ought to be at the top of the list of priorities of planning and development policy.

This process of integrating environmental concerns into development policy, known as mainstreaming, which in the European Commission Environmental Integration Handbook (2007) is referred to as "the process of systematically integrating a selected value/idea/theme into all domains of EC development cooperation to promote specific as well as general development outcomes", necessarily implies a shift in institutional culture and practices. Environmental considerations must be addressed both at the centre when long-term strategic planning is undertaken and at the periphery (local, companies and organisations). Environmental issues must be seen as an integral part of the decision-making process at every level, and not just a purely sectoral issue.

Moving on to talk about the drivers of this integration, principal among these have been the European Union (EU) and Organisation for Economic and Cooperation Development (OECD), whose work has been fundamental.

The EU's commitment to include environmental concerns in its development programmes and projects is an integral part of a broader commitment to sustainable development. Although earlier references to this integration process exist, such as the Treaty of Amsterdam (1997) and the Cardiff process, launched in 1998, it was the European Commission's Communication on Integrating environment and sustainable development into economic and development cooperation policy, published in May 2000, that made this process a reality.

In 2003, the Commission issued a Communication to the Council and the European Parliament on climate change in the context of development cooperation. This reinforced the need to include and integrate the problem of climate change into the strategic frameworks by concluding that concerns about climate change would be better addressed if integrated into other development cooperation policies, such as energy, transport, research and technology, water resource management and so on.

The links between development and the environment are also explicitly recognised in The European Consensus on Development (2006), a document that highlights the need for a reinforced approach that integrates cross-sectoral issues, making systematic and strategic use of all the resources available for that purpose.

It should be noted that the European Union as a whole is responsible for more than 50% of global development assistance and is a key partner in this process.

Portugal was a founding member of the OECD and has been an active participant in the organisation's activities for more than 40 years.

The 1st Joint High-Level Meeting of the OECD's Development Assistance Committee (DAC) and the Environment Policy Committee (EPOC) took place in 2006 and was a crucial step in this process. This was the first time that the OECD committee responsible for development aid policy had met at the highest level with another OECD sectoral committee (EPOC) and it represented a clear sign of the recognition by donor countries of the importance of integrating the environment into development aid policies.

The chief aim of this meeting was to establish the foundations for a strategic understanding between the environment and development cooperation committees. The meeting sought to identify key problems in the relationship between the environmental and development policy fields, specifically those stemming from the challenges of reducing poverty and of enhancing sustainable development, in an attempt to reach the development goals set out in the UN's Millennium Declaration. It also represented an opportunity to debate ways to improve coordination between multilateral and bilateral institutions that operate in the environment-development nexus.

The result of this meeting were the following two documents:

- The Framework for Common Action Around Shared Goals (2006). This Action Framework was intended to help partner countries define their own environment and development priorities, as well as integrate environmental considerations into policies and strategies for development and poverty reduction, in line with the principles and goals laid out in the Paris Declaration on Aid Effectiveness.

- The Declaration on Integrating Climate Change Adaptation into Development Cooperation, adopted by ministers, underlined the specific environmental problems faced by developing countries due to the effects of climate change and represented a very important tool in helping these countries, particularly with regard to adaptation.

A document from the same meeting—"Integrating Climate Change Adaptation into Development Cooperation"—was then published by the OECD (2009).

The 2nd High-Level Meeting on Environment and Development was attended by 75 high-level delegates, numbering 13 ministers, deputy ministers and agency heads, and representatives from all the OECD member states. This second meeting represented an excellent opportunity to take stock of the progress made since the joint ministerial meeting in April 2006.

The meeting also set out an ambitious agenda for establishing closer links between developing communities and the environment. Among the key issues of this agenda were:

- Sustainable financing for water supply and sanitation;
- Financing for climate change adaptation;
- Capacity-building for environmental governance and management in the context of the Accra Agenda for Action;
- Aid for low carbon development: approaches for simultaneously achieving the goals of mitigation and development.

For some time, therefore, we have been moving forward. But despite the distance travelled, in many cases the words of the 2008 Report of the Caribbean Natural Resources Institute still remain very pertinent: "Environmental problems only come to the fore when there is a crisis or a problem that affects a broad section of the population." Recently, we have witnessed the huge tragedy in Japan and closely followed how the country has struggled to contain a nuclear catastrophe. This disaster has once again brought the discussion about alternative sources of energy to the forefront.

According to scientific projections, we have very little time left to act, while future generations expect to inherit a better world than the one left to us by earlier generations.

In my final thoughts on this theme, I would like to look at Portugal and the steps it has taken to integrate the environment into development policy through strong and structured cooperation at the institutional level and by establishing relations with all social partners at the national and international level.

Portugal is a member of almost all the international fora dedicated to environmental issues and development, and also has special responsibilities in terms of development aid to nations with which it has close historical ties, such as the Community of Portuguese-speaking Countries (CPLP).

As we mentioned above, the degradation of the environment calls into question the ability to achieve the Millennium Development Goals. Protection of the environment, whether through specific projects, or via the integration of environmental concerns into other schemes and programmes, is therefore a necessity if sustainable development and poverty reduction are to be tackled. Following the recommendations of the main international environmental and development organisations, and in response to the specific needs identified by the countries in question, Portuguese cooperation has thus placed particular focus on protecting the environment as a means of fostering development.

Portuguese cooperation has been aimed at identifying the types of actions possible within the framework of the Fast Start scheme. Financing for these areas should cover the bolstering of mitigation measures, including the reduction in emissions from deforestation, the strengthening of adaptation measures, the reinforcing of development and technology transfer, and the boosting of institutional capacity-building.

Bearing in mind that the Millennium Development Goals are limited to a great extent by the effects of climate change, specifically due to its effects on human health, on the quality and quantity of water available, on agricultural production and on the threats to biodiversity, Portuguese cooperation follows the EU's guidance on integrating climate change into its development policy, thereby bolstering its activities in this area with different countries, insofar as the aid given for climate change adaptation is also, in itself, a form of development aid.

The cooperation actions set out aim to comply with Goal 7 of the Millennium Development Goals and the commitments of the Johannesburg Summit. As such, they introduce adequate management of environmental resources and, in particular, water resources; the protection of coastal zones and access to water and sanitation; and also introduce measures to combat global environmental threats caused by climate change and desertification.

While we are aware that Portugal's public development aid budget is under the EU minimum of 0.51%, we share the vision of the Portuguese foreign ministry to make the environment and climate change a central plank of cooperation strategy, despite the current economic situation.

In conclusion, the world is constantly facing new challenges that put our abilities to the test and underline our weaknesses. But the future is our responsibility.

Because of the effect it has on us, the environment demands that we act in a balanced, consistent and responsible manner. Our political decision-makers, however, often have to take controversial and difficult measures in an attempt to resolve the universal desire to satisfy needs. It is these needs that may have to be rethought given our over-consumption of the world's scarce resources.

We must always bear in mind the importance of the environment not just as a resource to be protected, but also for its potential contribution to a country's economic and social development.

How can this potential be exploited? What are the opportunities and obstacles for the environment in this context? These are the questions that have to be properly assessed and whatever development plan chosen will have to pay close attention to these strategic assessments. We need to know the best way to allocate scarce resources and to protect existing ones. The choices we make must allow the costs and benefits of policies to be compared with the costs of "political inertia".

Returning to Portugal's experience, we know that the renewable energy policy adopted will be a key tool in the country's economic recovery, since it helps to attain the goals of energy security, sustainability and competitiveness. Equally, it helps the country to fight climate change and to meet its commitments in this area. Targets must be realistic and accessible with regular and updated monitoring where necessary, providing the private sector with reassurance. Environmental assessments must continue to be conducted and broadened to include greater public scrutiny, incentivising the participation of one and all.

It is also crucial to engage the public and to create more effective incentives for the active involvement of the private sector in the environment. Reform of the tax structure will be key to this and will contribute to fostering the right behaviour by all social partners, from the public to private sector, and from producers to consumers.

We also speak nowadays of the green economy, and of the need to bolster sustainable development in the face of coming adversities. I believe that all of the work undertaken in recent years will allow us to lay the foundations for the green growth that we must develop. It cannot be postponed or put on stand-by. The work done so far has to continue, both in terms of the environment and development cooperation, placing the focus on policy initiatives and strategies that promote eco-innovation and sustainable models of production and consumption.

References

"Ambiente, Economia e Empresa", SAER, Maio 2010

European Commission (COM (1998) 333) Communication from the Commission to the European Council of 27 May 1998 on a partnership for integration: a strategy for integrating the environment into EU policies (Cardiff—June 1998)

European Commission (COM (2000) 264) Communication on Integrating environment and sustainable development into economic and development cooperation policy—elements of a comprehensive strategy

European Commission (2003) Climate change in the context of development co-operation. Brussels

European Commission (COM (2006)/C46/01) Joint statement by the Council and the representatives of the governments of the Member States meeting within the Council, the European Parliament and the Commission on European Union Development Policy: 'The European Consensus'

European Commission (2007) Environmental Integration Handbook for European Commission Development Co-operation

OECD (2006) Framework for common action around shared goals

OECD (2009) Integrating climate change adaptation into development co-operation. OECD, Paris. http://www.oecd.org/env/adaptation/guidance

OECD (2010) OECD development assistance committee, Peer Review 2010. OECD, Portugal

OECD (2011) OECD environmental performance reviews. OECD, Portugal

UNEP (2010) UNEP Annual Report 2010

Chapter 13
The World We Want and the World We Have

Who Ties the Knot?

Zenita C. Guenther

Abstract The notions we present in this paper were meant to provoke open discussion, and work through different points of view on how to deal with our often troublesome modern world. The paper would be composed after considering whatever should come out of the discussion. However there was no opportunity for discussing it at the meeting, so we are still working from a one-sided point of view, initially chosen to stimulate different reactions. If anyone gets to read this paper I foresee a probability of some "head shaking" as was intended by the author. If so occurs, please drop me a line, and I will be more than happy to listen to whatever you want to say, add, inform or contradict, and I am sure that new ideas for action may still develop.

13.1 Our World

When taking on the task of thinking about the World we have, as compared with the World we want, some unavoidable ideas come to mind: looking around we see this group of selected persons from many different places and different fields of activities, making themselves available to examine our world, the planet we live in, the acknowledged mistakes we did in the past… still happening in the present… with probability of repeating them over and over into the future… and what do we see?

The Earth, this (small) Planet where we live
We here had a look at the *Biological Diversity*, were put face to face with what is happening concerning the—*Water Challenges of the twenty-first Century*, and heard some evidence of at least one consequence on—*Confirmed Climate Changes*. There is no doubt that our proud small planet is meeting with situations that open ways to not small problems for not a far future….

The people…
We seem to be more than lightly concerned with human aspects related to these problems, such as—*Mental Health Effects of the Environment*, what touches directly

Z. C. Guenther (✉)
Centro para o Desenvolvimento do Potencial e Talento, UFLA, Lavras, Brazil
e-mail: zeguen@def.ufla.br

on—*Human Rights and Harm*, we are also somewhat surprised that people are not aware of—*What Environmental History Teaches*, although showing a few attempts to move toward desired changes, such as the—*Integrated Approach to Conserve Mangrove Forest at Keti Bunder*, without ignoring whichever new technologies we can gather, as pointed out in—*The Contribution of the Barcode of Life Initiative to the Discovery and Monitoring of Biodiversity*.

And people elsewhere…
Opposing to what is being experienced in this conference room, most people elsewhere in our same small planet are forced to live with undesirable situations over and over, repeatedly showing the same sad pictures of sickness… ignorance… poverty… loneliness… violence! That is the world we definitely do not want, unfortunately have not as yet been able to change, and obviously cannot ignore forever….

13.2   The World We Do Not Want!

Our modern world takes pride on its advances in science and technology, mostly because they brought considerable improvements and advances into our lives; it is not hard to observe how many people, communities, even whole countries have reached high levels of material comfort, quality education, long and healthy lives for themselves and their descendants; and some common people are close to a number of individuals so unbelievably rich that it is hard for them to imagine a life surrounded by everyday poverty, sickness, ignorance….

Without going much further, it is obvious that one can partially agree with material accomplishments as some measure of success in terms of wining the struggle to be alive…. But only partially, because it is also undeniable that the same countries seen as developed, well fed, cared and educated, submit people to a whole new set of values and conditions that hit hard on their everyday lives. Looking at mankind as more than a simple animal species one wonders if unlimited and generalized material wealth is in fact a path to peace and happiness for the world. Let's browse over same points that could be considered:

- **The built in violence within the economics power system…**
  The stronger and wealthier a group of people gets the more they seem to be faced with quantitative external values and possession of material goods superimposing upon qualitative values and internal conditions inherent to their lives… to the point that practically all they do, say, think… or do not do, say or think… comes up anchored on materialistic basis, and all of the possible ways to move ahead that are worth considering seems to be toward one and only direction: *Money solves all problems in life… and in the world!!!* But… we still have all those problems mentioned above, and when a community of any profile reaches the desired level of material wealth to be considered "developed", contrary to what should be expected, such problems do not get solved, but are usually aggravated.

- **The ridiculous of Media Imperialism...**

  It is a "no secret saying" that "...*if a medical doctor thinks he is God, a journalist is sure that he is!*" (R. Noblat, journalist). And it does not apply only for what is in the local newspaper. In fact no one gets to be heard on anything, except when what he wants to say is already "*published*", was on TV, or is in the *internet*.

  So, if whatever you have to say that is not already in print is held out as doubtable, it follows that whatever comes in print is to be believed... In academic environments there is some gossiping around the question "*Could God be a University Professor?*" raised and examined by a group of "respectable personalities", experts on the ritual involved with high level teaching. The results were not favorable to God, mainly because he wrote only one book, with no bibliographical references, never published anything in scientific journals, and apparently had unacknowledged co-authors working on his book ....

  Surely it does not sound serious? Well... let's not forget how often printed material sounds utterly ridiculous, and yes sir, I do have references taken from first page headlines: *Miners Refuse to Work after Death* ...(Did they get way with that?); *If Strike Isn't Settled Quickly, It May last a While*...(A plausible conclusion.); *Cold Wave Linked to Temperatures*...(Hum...why not to thermometers?); *Typhoon Rips Through Cemetery: Hundreds Dead*...(Good choice!); *Red Tape Holds up New Bridges*...(And innocent passersby never know these things...); *Police Begin Campaign to Run Down Jaywalkers*...(That will teach them!); *Hospitals are Sued by 7 Foot Doctors*...(Who hired 10 foot lawyers...); *Teacher Strikes Idle Kids*...(And they well deserve it!); *Kids Make Nutritious Snacks*...(Anyone for tasting?); *Local High School Dropouts Cut in Half* ...(One of the parts could go to classes...). And that shall close my case on the credibility of the media....

- **The inescapable cultural sicknesses**

  There are signs that the culture we are building into our everyday lives is unhealthy. That does not refer to the often mentioned new forms of mental and emotional illnesses sprouting everywhere at the individual level, but to a deep sickness built into the way we live, raise our children and educate them:

  - **Insatiable consumption**—One of the most salient characteristics of our modern day world, possibly more visible in the Western Culture, but spreading fast to nearly all countries in the world, is the hunger for consuming everything that is available, or heard of, be it goods or services. So far it is not clear if the self-perpetuating consumption needs are created by the marketing, or if in fact they made room for expanding the field for marketing. What remains as a fact is that people expends unaccountable amounts of money, and mortgage way whole parts of their productive lives, buying what they do not need with money they do not have.
  - **Shallowness of Commitment**—The fast speed of change invading every aspect of human life do not allow time or room for people to really adopt a point of view, take a personal standing or make any kind of commitment

regarding most of the situations they have to face. As a result commitments are shallow and not really binding, even on what could be considered relevant parts of one's life, such as marriage, family, group allegiances, or one's chosen profession. There is a floating feeling that it is dangerous "to care" because by caring one is open "to get hurt" somehow.

- **Ideological weakness**—With no deep level commitment it is hard to have strong ideological principles related to being a part of a group of people born and living together. There seems to be no point to occupy one's mind with debating or choosing a position regarding ideals and directions to be followed by groups or societies. Therefore people do not care, do not get involved, and do not feel motivated to embrace long term goals or to set basic principles for their groups, cities or nations.
- **Political hypocrisy**—Of course nations have to be run and official governments need to be set. Since most of them are elected, there has to be someone interested in political matters… and someone certainly will be, since we hear of no shortage of candidates to political offices. On the contrary, quite often there is an abundance of them with little difference between each other… But by being raised in an environment of ideological weakness, most of them will not have very strong beliefs regarding the nation's principles… Could that be an open door to the well-known political hypocrisy that plagues most governments everywhere?….

13.3   How People Manage to Survive?

Looking at the kind of world here described, although aware that we do not want it, allows predicting not an easy and happy life for *normal human beings*. And sure enough, there are signs gathered through data collected mostly on "developed nations", showing unhealthy ways to look at self and others:

- **Learned helplessness**—In most situations people feel impotent and helpless to attempt doing anything other than whatever is already established; when facing any of the everyday problems he encounters, a person is unclear about where to go, if anywhere, and what to do, if anything… so he does nothing! The common man on the street is sure that he can't solve anything, that "*It is no use to bother…*" or "*One cannot fight city hall….*"
- **Unquestioned obedience to authority**—In any case, the mass of survivors usually do as they are told, do not ask and do not tell, provided that the order comes from someone or somewhere above. In other words, it is quite clear that "*Those who can… give the orders, and those who 'know better' … obey them!*"
- **Conformity to group pressure**—When not obeying orders or following directions from higher hierarchies people survive by looking around and doing what others are doing, for one of two reasons: he does not in fact care, and just follows the crowd; or he cares and feels bad about doing that, but somewhere some group has the power to make him do it… "*or else….*"

- **Diffusion of responsibility**—A protective umbrella to hide people from the need to be held accountable for any shaking decision is provided by the general impersonal situation that no one is responsible for whatever is decided. That is to say that the individual never has any measure of accountability, since in any given situation the final words come to *"the committee has decided..."* or *"the majority chose...."*
- *Micro-ethics: the "more or less" principle*—Actually our modern man does not appear to recognize right from wrong in situations involving each other, group interaction and inter- relationships. The *micro-ethics* can help a person to reach some degree of peace with himself, by making decisions based on the *"more or less"* principle: one cannot act too wrong by being *more or less honest, more or less concerned, more or less responsible, more or less involved, more or less productive....*

Bottom line: by trying to survive in the world which we do not want, human beings who are genetically planned and born as unique and practically not repeatable in their individuality, are well in their way to a high level of general "de-individualization" (Maslow 1962).

13.4   Bridging the Gaps

Some of the acknowledged research on the humanistic movement (such as can be seen in works by Maslow, Combs, Bugenthal, Antipoff, Freire Maslow 1962; Combs and Snygg 1959; Antipoff 1992) looks for what can be understood as essentially *"the nature of man"*. Their findings show a picture of people, as human beings, invariably reaching out for:

- **Meaning, sense of value in one's own existence**—most humanists point to the search for meaning in one's own life as essentially a natural characteristic of humanity; it also follows that a large part of emotional problems, psychosomatic disorders, drug addition, urban violence and personal unhappiness are rooted in some vague sense of not finding meaning and value in one's own life.
- **Balance, harmony, internal and external consonance**—one area of deep needs in human beings is to reach an acceptable degree of coherency between one's attitudes, values and behavior, in all relevant aspects of his personal, family and community life.
- **Control over one's own life**—one has to have assurance of enough latitude, time and opportunity to make and hold decisions regarding his personal and family life.
- **Peaceful, sensible living together**—indications are that people of all age do wish to live in harmony, peace and togetherness with each other, in all walks of life, what is contrary to the observable direction in modern ways: a desire for totally independent, although alienated and lonely living styles for individuals.

- **Self-esteem and esteem by others**—and more than ever, the lonelier a person gets the stronger is the need to be assured of holding a high level of self-respect and self-esteem, all properly corroborate by others' respect and esteem for him.
- **Making a significant contribution in some area**—it can be said that this is the great dream of modern man during his lifetime: to do something big, good and powerfully important, something that will mark his life and others, now and for the future.

So, the human kind seems to be profoundly concerned with many themes of day by day living not included in the concrete field of material comfort and worldly possessions (Rogers 1961).

Duality and contradictions However, when talking about "*human nature*" as we are trying to do, one cannot escape from considering another side expressing the uncertainty of a dual nature, apparently more visible to artists, "*I'm neither good nor bad... I'm human and I'm sad.*" (Olavo Bilac, poet); or as portrayed by the novel writer Kundera"...*The crew that runs my soul does not get along with the crew that runs my body.*"

And so are human beings... believed by themselves to be the only animal species which is able to ask questions such as "*Who am I?... Where do I come from?... Where do I go to?*", and to attempt some kind of an answer to them.

... Optimistic assumption Humanistic psychologists do not hesitate to proclaim that human beings are by nature oriented to the positive side of life, and in *normal conditions*, a person tends to grow, develop and improve his own perceived self. On the same line of thinking, Carl Rogers (Rogers 1961) is convincing when presenting the most basic tendency in life—human life or otherwise—as a built in determination to move away from harm and toward higher levels of adequacy, as individuals, groups, and mankind as a whole.

... Discouraging observations Nevertheless even thinkers like Abe Maslow (psychologist), considered one of the good humanistic minds of our times, could not deny puzzling over the fact that "*most people in power are often bastards*"... But the question he struggles with refers to clarify if is "*to have power that exposes the bastardy that men always had, or more bastards get to higher levels of power than other people...*" (Maslow, in Memorian Volume 1972). Rui Barbosa (lawyer and writer) shows his disappointment and bitterness in observing that "*From seeing the triumph of nullities... the prosperity of the dishonorable... the growth of injustice... the maintenance of power in the hands of the bad... Man reaches the point of being discouraged with virtue, looking down at honor and is ashamed of being honest.*"

... Conundrums... Of course there are some practical questions muddling around: "*Is it better to cooperate, or to compete?*", or when competing "*if you cannot beat them... join them...*" What is really important to change in a person: their way of thinking, which is to say their "*attitudes*", or their way of doing, that is their "*behavior*"? And isn't possible that "*...if you cannot do what you think is correct... you start thinking that what you are doing is correct ...*" Or, as expressed by the King of Siam, "*It surely is such puzzlement!!!*"

13.5 What Can Be Done?

All clues point toward Education… According to the pedagogical discourse, Education is to be in full charge of developing and perfecting human beings in many dimensions of their lives: in their ways of being, acting and behaving; their personal goals and expectations; their give-and-take relationships within the physical and social world; their inter-relationships with other human beings, alike or different from themselves; and to build good intra-relationships with their own interior selves. In few words, said by Paulo Freire: "*Education is to improve myself with the others with the world*" (Freire 2005).

- **We believe in Education**—and that was witnessed in this meeting by Joan Freeman (UK) showing the relevance to gather "*Clever Minds for Climate Control*"; Michael Marzolla (USA) documenting his experience with "*Participatory Development of Fotonovelas for Environmental Education and Community Empowerment*"; Laura Barraza (Mexico) presenting her "*Pedagogic Strategies on Teaching Science and Environmental Education*"; Mario Freitas (Portugal) bringing on the idea of "*Public Participation as Leverage for Building Up more Sustainable Societies*"; and Elizabeth Silva (also Portugal) pointing out the role of "*Education for Sustainable Development*".
  Not only here but elsewhere good educators like Helene Antipoff (Brazil) go as far as declaring that "…*true social, economical, political and spiritual progress will not be accomplished except through education… Therefore it is due to educators the maximum accountability for both: The evil which affects the peoples, as well as the goodness and well-being that is afforded to them*" (Helena Antipoff 1970).
- **And give tips to Educators**—there are many suggestions in a variety of ideas to better education, coming from thinkers everywhere, as for instance Edgar Morin describing some lines of *knowledge* we need to provide in the twenty-first century: to "know" without error or delusion; to acquire "pertinent knowledge"; to learn our own "Human Identity"; to comprehend each other as human beings; to learn how to deal and acknowledge "uncertainties"; and to develop awareness of our "planetary condition" in the universe.…

13.6 Who Will Do It All?

Can schools truly and really accomplish all of it… or there is room for doubt…

It is a fact that teachers everywhere struggle to be faithful to the ideals presented by a nice, hopeful, well put together educational discourse, which keeps changing to include new fancy words, but remains passing on virtually the same old messages, year after year… But it is also true that the day by day living within most schools is heavily weighted with contradictions, conflicts, incoherencies… and long days dragging on endlessly through a tedious tiresome unchangeable routine, altogether different in thinking and doing from what is stated as goals and purposes of Education.

For a start, Education is said to aim at preparing youngsters for life by teaching them to be *autonomous, dependable, responsible, committed, able to think, competent, creative, sharing and cooperating,...* what shows up in any school curricula. But *real life* all around the schools is going on within the frameworks of a social system that is industrialized, mechanized, synchronized, pattern oriented, hierarchically set, centralized, repetitive, routine bound... *and so are the schools*, by its own organizational structures....

With such characteristics in the social environment, the so called *real life* imposes real demands in and out of schools for behavioral patterns leading to synchrony, submission, competition, hierarchy, obedience, homogeneity, resistance to tedium, routine repetition... all contrary to the stated educational purposes. Trapped by the impossible contradiction of opposed goals and mixed demands, schools had to face the futility of its own existence. And survives by building its functioning basis within a *hidden curriculum*, in which students and personnel alike are subject to everyday life experiences that are the opposite of what is intentionally taught in the classrooms! All in all it kills any hope to improve human life through Education.

Not only that, but such a phenomenon makes it possible to argue that the schools, as social institutions, are doing precisely what they are meant to do: nourishing and maintain the social system, with not much concern for bettering the world or improving quality of human life....

If these ideas are not disputed, it cannot be expected that the needed *teaching and learning* for the world we want would happen by "working with the schools".

If not schools, then... what? Of course there are supplementary questions, such as "why developed and 'educated' countries" pollute more, create more waste, expend more resources and demand more living space than "not developed uneducated countries"?

In this meeting we have seen some outlines for other solutions, and some speakers suggesting that the changes should be implemented by force of law and public policy. That may be so, but the clear implication that there is a kind of *education* that comes faster through "*making them do it, and punish if they don't*", is of course in line with de-individualization principles. Considering that "punishing" usually means a financial penalty, such as a fine, or paying for something supposedly needed elsewhere, here comes another sign of cultural sickness: "*Money solves all problems*". Does it not vaguely sound like a sick answer for a sick world?

Another track of thought goes toward implementing education, but not in schools. This is a new effort to raise faith in Education by means of informal teaching and learning (Illich 1971). We cannot ignore that there may be hope in this idea, if and when we cross the barrier of financing such initiatives. Can we trust each other enough to let people try their ideas at face value, and wait for long term results? How many of them will have strength to endure the needed time dimension for building credibility? And how much public money are we willing to risk on the diversity of ideas that can be generated if resources are made available?

It all hits hard on the question of deciding the future of Schools. Some futurologists, playing simultaneously with fact and fiction in our world, show great hopes for a bright life-around-the-clock for people in the future (Illich 1971). Education is included in their visions, by proclaiming that the 24 hour-a-day society will challenge the "industrialized organization" forced upon schools (Toffler 1981).

On the other hand, it is not hard to notice the effort from modern pedagogy to spread the need for individualization in everyday life by demanding undisputable respect for personal and group rhythms, interests, abilities and styles of life chosen by the individual. That applies to all pupils of all ages, and goes without mentioning the growing preference for informal and out of school education in all walks of life. Would that idea look pretty much like another "sign of the times…"?

References

Some Traditional Still Provocative Readings

Antipoff H (1970) Como aprender a ser feliz- That is the question. Caderno Feminino do Jornal Estado de Minas, Maio

Antipoff H (1992) A Educação do Bem Dotado, Coletânea de Escritos de Helena Antipoff, Vol V. Rio de Janeiro, SENAI

Combs A, Snygg D (1959) Individual behavior: a perceptual approach to behavior. Harper & Row, New York

Freire P (2005) Pedagogia do Oprimido, 45a. Edição, Ed Paz e Terra, Brasil

Illich I (1971) Deschooling society. Harper & Row, New York

Maslow A (1962) Toward a psychology of being. D Van Nostrand, Princeton

Maslow A (1971) The farther reaches of human nature. The Viking Press, New York

Maslow A (1972) Some basic propositions of a growth and self actualization psychology. In: Combs A (ed) Perceiving, behaving, becoming—a new focus for education. NEA, Washington (ASCD, 1962 Yearbook)

Rogers C (1961) On becoming a person: a therapist's view of psychotheraphy. Houghton Miffin, Boston. http://erclk.about.com/?zi=12/l69

Toffler A (1981) The third wave. Pan Books, London

Index

Lightning Source UK Ltd.
Milton Keynes UK
UKHW021840260321
381059UK00002B/2